Penguin Education

Penguin Library of Physical Sciences
Elementary Particles
I. S. Hughes

Advisory Editor
V. S. Griffiths

General Editors
Physics: N. Feather
Physical Chemistry: W. H. Lee
Inorganic Chemistry: A. K. Holliday
Organic Chemistry: G. H. Williams

Elementary Particles

I. S. Hughes

Penguin Books

Penguin Books Ltd, Harmondsworth, Middlesex, England
Penguin Books Inc., 7110 Ambassador Road, Baltimore, Md 21207, USA
Penguin Books Australia Ltd, Ringwood, Victoria, Australia

First published 1972
Copyright © I. S. Hughes, 1972

Filmset in Monophoto Times by
Oliver Burridge Filmsetting Ltd, Crawley, England
and made and printed in Great Britain by
Compton Printing Ltd, London and Aylesbury

There are therefore Agents in Nature able to make the Particles of Bodies stick together by very strong Attractions. And it is the Business of experimental Philosophy to find them out.

Newton, *Opticks*, p. 394

Contents

Editorial Foreword

For many years, now, the teaching of physics at the first-degree level has posed a problem of organization and selection of material of ever-increasing difficulty. From the teacher's point of view, to pay scant attention to the groundwork is patently to court disaster; from the student's, to be denied the excitement of a journey to the frontiers of knowledge is to be denied his birthright. The remedy is not easy to come by. Certainly, the physics section of the Penguin Library of Physical Sciences does not claim to provide any ready-made solution of the problem. What it is designed to do, instead, is to bring together a collection of compact texts, written by teachers of wide experience, around which undergraduate courses of a 'modern', even of an adventurous, character may be built.

The texts are organized generally at three levels of treatment, corresponding to the three years of an honours curriculum, but there is nothing sacrosanct in this classification. Very probably, most teachers will regard all the first-year topics as obligatory in any course, but, in respect of the others, many patterns of interweaving may commend themselves, and prove equally valid in practice. The list of projected third-year titles is necessarily the longest of the three, and the invitation to discriminating choice is wider, but even here care has been taken to avoid, as far as possible, the post-graduate monograph. The series as a whole (some five first-year, six second-year and fourteen third-year titles) is directed primarily to the undergraduate; it is designed to help the teacher to resist the temptation to overload his course, either with the outmoded legacies of the nineteenth century, or with the more speculative digressions of the twentieth. It is expository, only: it does not attempt to provide either student or teacher with exercises for his tutorial classes, or with mass-produced questions for examinations. Important as this provision may be, responsibility for it must surely lie ultimately with the teacher: he alone knows the precise needs of his students – as they change from year to year.

Within the broad framework of the series, individual authors have rightly regarded themselves as free to adopt a personal approach to the choice and presentation of subject matter. To impose a rigid conformity on a writer is to dull the impact of the written word. This general licence has been extended even to the matter of units. There is much to be said, in theory, in favour of a single system of units of measurement – and it has not been overlooked that national policy in advanced countries is moving rapidly towards uniformity under the *Système International* (SI units) – but fluency in the use of many systems is not to be despised: indeed, its acquisition may further, rather than retard, the physicist's education.

A general editor's foreword, almost by definition, is first written when the series for which he is responsible is more nearly complete in his imagination (or the publisher's) than as a row of books on his bookshelf. As these words are penned, that is the nature of the relevant situation: hope has inspired the present tense, in what has just been written, when the future would have been the more realistic. Optimism is the one attitude that a general editor must never disown!

N. F.

Preface

This book is intended for undergraduates or others coming to the subject of particle physics for the first time. For this reason the only prior knowledge assumed is of the elements of quantum theory and statistical mechanics.

The story of the development of elementary-particle physics in the years since the war has been one of continuous excitement. Much of this has been due to an unceasing interplay of experiment and theory in the best classical tradition. Scarcely a year has passed without a remarkable advance in theory or experiment, such as the discovery of the antiproton, of K-mesons and of hyperons; the Gell-Mann–Nishijima scheme and strangeness; parity non-conservation; two neutrinos; the still-growing list of short-lived resonances, the SU(3) symmetry scheme and the Ω-particle; the quark proposal, and a good many others.

This rapid progress has been a consequence of, and a justification for, a parallel progress in technology and instrumentation. In the first chapter of this book I have attempted to outline the principal experimental techniques used in this work. I hope that this will enable the student to visualize how the many experiments referred to in later parts of the book have been actually carried out, since such an understanding is necessary to a proper appreciation of the subject. While I have not adopted an historical approach, I have felt it desirable to discuss the way in which many problems were originally seen and subsequently solved: as for instance the puzzle of the muons when first observed, the τ–θ problem and others, since the solution of the problems is itself instructive and aids the understanding of these phenomena.

In treating the theoretical aspects of the subject my approach has been to deal only with the theory as derivable from conservation laws, since the dynamical theory is both more difficult and less secure.

Particle physics is currently a very active subject in its theoretical and experimental aspects, and in the building of machines such as the new accelerators, the very large bubble chambers and new spark-chamber devices. I have tried to bring the discussion as near as possible to current work and in doing so I take the risk that some of the most recent results and ideas may prove in time to be wrong. I believe that this risk is justified by the attempt to show that the subject is very much alive and is continuously throwing up the most fundamental and challenging problems.

The subject has been presented in sufficient depth to give the student an understanding of its fundamental nature, its fascination and its recent startling progress. In an introductory text, however, many subjects have to be dealt with superficially or not at all and some results presented on trust. In a subject as active as particle physics it is not entirely straightforward to recommend books for further reading, since much of the most useful material is contained in reports of summer schools such as the excellent series organized by CERN, the notes for which are published as CERN yellow reports. However, as a thorough background text on the theory I recommend *The Physics of Elementary Particles* by H. Muirhead. A good introduction to group theory is provided in *Lie Groups for Pedestrians* by H. J. Lipkin. The subject of relativistic kinematics is dealt with in more detail and also more extensively than is possible in this book in *Relativistic Kinematics* by R. Hagedorn.

The increasing use of the standard SI units in all fields of physics is clearly generally to be welcomed. However, in certain fields the magnitude of the quantities involved is such that it sometimes makes good sense to use units chosen for their particular convenience in the application in question. It is also important that texts should be in line with actual usage in an active field of science. Although it is clear that there is a two-way interaction between textbooks and practice in the field, the transition from one system to another cannot in fact be a sharp one. For these reasons it will be found that in this book I have not attempted to conform exclusively to SI units.

I am very much indebted to Dr W. T. Morton for his careful and critical reading of this book, as a result of which many improvements have been made in the text and many errors corrected, and also to Dr W. M. Gibson for a number of valuable comments on the final draft. I am grateful to the authors who have permitted me to reproduce figures from their published papers, to the CERN photographic service for providing certain photographs, and to the following journals for permission to reproduce diagrams originally published therein:
Nature, Nuovo Cimento, Philosophical Magazine, Physical Review, Physical Review Letters, Physics Letters.

Finally it is a pleasure to acknowledge the work done by my wife in typing a difficult manuscript.

Chapter 1
Particle beams and particle detectors

1.1 Introduction

An important part of the study of particle physics is an understanding of the ways in which particles are detected, their properties measured and their trajectories controlled. There exist a number of types of detector and methods of handling particle beams which are very commonly used. This chapter attempts to outline the main features of these techniques. No more technical detail is included than is essential to an understanding of the uses of these techniques in the study of particle physics. In the chapters which follow we shall assume that these techniques are familiar to the student, so that it will not be necessary to describe in detail the technique used in a particular measurement if it is of a standard kind.

1.2 Particle beams: introduction

In carrying out experiments in particle physics, it is generally necessary to know the nature and momentum of the incident particles which interact with targets of hydrogen or deuterium (normally liquid targets) or complex nuclei. This may be achieved in counter and spark-chamber experiments either by using a pure beam, in which the wanted particles are present in a narrow momentum bite with little background, or by using an impure beam, but identifying the particles before they pass into the target, and studying only processes which occur in coincidence with a wanted particle in the target.

Bubble chambers cannot be triggered and their cycle time is relatively long (~ 0.2–2 s) so that for bubble-chamber experiments it is generally best to use as pure a beam as possible. In a collision of protons of greater than a few thousand million electronvolts with a solid target in an accelerator many kinds of particle will be emitted. With positive charge we expect p, π^+, K^+, Σ^+, and due to decay of these μ^+ and e^+. The Σ^+ quickly decay but an unseparated positive beam will contain all the others. With negative charge we will get $\bar{p}$, π^-, K^-, Σ^-, Ξ^-, μ^-, e^- of which only the hyperons have very short lives. The ratio of protons and π^+ to K^+, or of π^- to $\bar{p}$ and K^- at the accelerator target is always high (from 20 to $> 10^3$ depending on the momenta of the incident and produced particles and on production angle), so that if a beam of kaons or antiprotons is required a very efficient system for rejection of protons and pions is needed.

In the following discussion only the main features of particle beams will be discussed. In practice the aberrations of the beam components are important in the production of good images for particle separation. Those wishing to pursue the subject in more detail are referred to Chamberlain (1960), Sternheimer and Cork (1963) and Sandweiss (1967).

1.3 Magnetic devices

One of the first elements of every particle beam is a *bending magnet*. The particles pass through a uniform magnetic field and suffer bending according to

$$p = \frac{\rho e H}{c},$$

where H is the magnetic-flux density, ρ the radius of curvature, e the charge on the particle, c the velocity of light and p the momentum of the particle. A slit after the bending magnet can thus select the desired momentum bite. Within the bite there will be dispersion which may be removed by a compensatory bend in the opposite

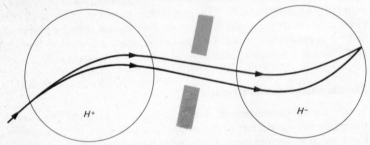

Figure 1 The use of two magnets with opposite fields normal to the paper to produce momentum analysis, using a collimator, with correction of the dispersion

direction (Figure 1). After such magnetic analysis we have a beam of well-defined momentum but still containing particles of different mass with different velocities.

Non-uniform magnetic fields, usually produced by *quadrupoles*, are employed to achieve focusing of the beam. While the bending magnets are equivalent to optical prisms, the quadrupoles are equivalent to lenses. A cross-section through a quadrupole magnet is shown in Figure 2. It is clear that on-axis particles are unaffected by such a magnet. In the plane AB, off-axis particles are deflected back towards the axis so that in this plane the quadrupole acts as a convergent lens. In the orthogonal plane, however, particles are deflected off-axis and the equivalent lens is divergent. Only with two or more quadrupole magnets can convergence in both planes be achieved.

1.4 Electrostatic separators

Magnetic analysis can yield a beam of well-defined momentum. The particles of different mass in such a beam will have different velocities. The electrostatic separator provides a means of separating particles of different velocities.

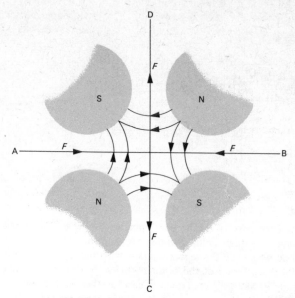

Figure 2 Diagrammatic cross-section through a quadrupole magnet.
The field is as shown. Negative particles passing normally into the plane of the paper
from above will suffer forces F as shown. Thus the quadrupole focuses such particles
in the plane AB and defocuses in the plane CD

Such a separator consists of a pair of parallel plates maintained at a high potential difference. Particles passing into the gap between the plates then experience a
transverse force Ee, where E is the field and e the charge. If the particle velocity is
v and the length of the separator is L the particle will receive a transverse momentum component

$$p_\mathrm{T} = \frac{EeL}{v},$$

as a result of passing through the separator. The angular deflection will be

$$\theta = \frac{p_\mathrm{T}}{p} = \frac{EeL}{p\beta} \quad (\beta = v/c, \text{ and we take } c = 1),$$

and the difference in angular deflection (Figure 3) for particles with different velocities proportional to β_1 and β_2 is

$$\Delta\theta = \frac{EeL}{p}\left[\frac{1}{\beta_1} - \frac{1}{\beta_2}\right].$$

Writing $\qquad \beta_1^2 = \dfrac{p^2}{m_1^2 + p^2}$

we have $\quad \dfrac{1}{\beta_1} - \dfrac{1}{\beta_2} = \left[1 + \dfrac{m_1^2}{p^2}\right]^{\frac{1}{2}} - \left[1 + \dfrac{m_2^2}{p^2}\right]^{\frac{1}{2}},$

which for relativistic particles is given by

$$\frac{1}{\beta_1} - \frac{1}{\beta_2} \simeq \frac{1}{2}\left[\frac{m_1^2}{p^2} - \frac{m_2^2}{p^2}\right].$$

So that $\quad \Delta\theta \simeq \dfrac{EeL}{2p^3}\Delta(m^2).$ **1.1**

Much effort has been devoted to the development of electrostatic separators which can maintain very high fields without breakdown. Unlike some other situations it is not possible to fill the separating tank with high-pressure gas as this would cause multiple scattering of the beam particles. Nevertheless separators capable of sustaining 10 MV m^{-1} over a $0\cdot1$ m gap have been developed.

The deflection of the wanted particles is generally compensated by vertical bending magnets at the ends of the separators.

The beam may be parallel or diverging through the separator. A typical situation is illustrated in Figure 3. The angular separation achieved by the separator is trans-

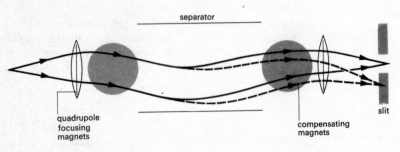

separator

quadrupole
focusing
magnets

compensating
magnets

slit

Figure 3

formed into a spatial separation at the 'mass slit' by a converging quadrupole lens, focusing at the slit. The wanted particles are then allowed to pass through the slit, while the unwanted particles are stopped.

Frequently two stages of electrostatic separation are employed with an intermediate bending magnet to get rid of off-momentum contaminants such as the products of beam particles which have decayed in flight.

It is possible to scan the images of wanted and unwanted particles across the mass slit by varying the current in the compensating magnet following the separator and detecting the particles with a scintillation counter behind the slit. The result of such a scan is shown in Figure 4, which illustrates the excellent separation of

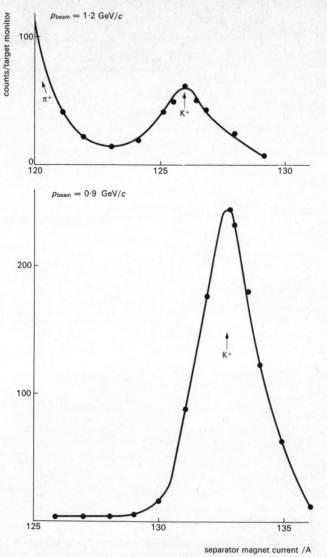

Figure 4 The result of scanning the particle images produced by an electrostatic separator across the 'mass slit' for two incident momenta. The K⁺ peak is very clearly seen separated from the π⁺. This result is from a low-energy separated beam at CERN

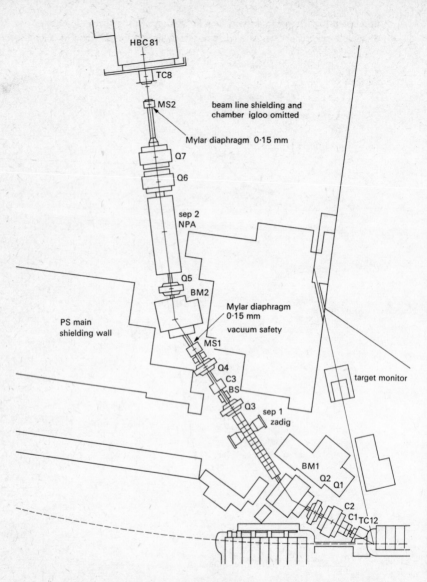

Figure 5 The layout of a low-energy separated beam from the proton synchroton at CERN for the 0·81 m hydrogen chamber. The Qs represent quadrupole magnets, BM bending magnets, sep. electrostatic separators, C and S collimators and slits.

K-mesons from π-mesons which can be achieved by this technique at momenta of a few GeV/c. The layout of a typical low-energy separated beam is shown in Figure 5, while the beam profiles in the horizontal and vertical planes are shown in Figure 6.

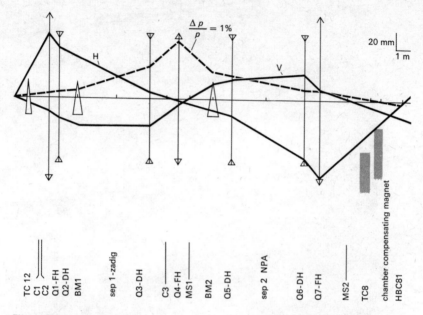

Figure 6 The horizontal (H) and vertical (V) beam profiles for the beam shown in Figure 5. The transverse scale is magnified by 50 compared with distances along the beam. The diagram shows the effects of the quadrupole magnets. The dashed line is for a particle with momentum 1 per cent off the design momentum

A photograph of a section of a separated beam layout at CERN is shown in Figure 7.

1.5 Radio-frequency separators

Equation **1.1** shows that for a fixed Δm the separation which can be achieved varies at the inverse cube of the momentum. The degree to which this problem can be compensated for at high momenta is limited, since E cannot be increased for practical reasons and, of course, an increase in L results in a greater number of decays in flight with resulting decrease in flux and increase in background. The highest momentum achieved in such a beam has been for 6 GeV/c K$^-$ mesons at CERN.

Figure 7 Part of a beam at the CERN international laboratory.
Parts of three horizontal bending magnets can be seen in the foreground
followed by a focusing quadrupole, another bending magnet and another quadrupole.
Following this is a vertical bending magnet associated with the 10 m long
electrostatic separator in the cylindrical tank (see Figure 3). The high-voltage feeder
for the separator can be seen entering the top of the tank. The beam passes on down
the vacuum tube to a bubble chamber beyond the shielding wall
(Photograph by courtesy of Photo CERN)

For beams of higher momentum a method of velocity separation has been developed depending on the time of flight of the particles between two radio-frequency cavities.

The principle of operation is illustrated in Figure 8. A bunch of particles of well-defined momentum passes through the first radio-frequency (r.f.) separator R1,

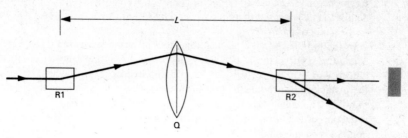

Figure 8

which is normally a cylindrical iris-loaded waveguide, and suffers a transverse deflection which will vary in magnitude according to the part of the r.f. cycle at which the bunch passes through the cavity. The deflected beam is focused on the second cavity R2 by a quadrupole system Q. If the distance L and the relative phase of the cavities is suitably adjusted the unwanted particles can receive a deflection exactly cancelling the original one, whilst that of the wanted particles is doubled. The unwanted particles are then stopped by a beam stopper S. The condition for this is

$$\frac{L}{\beta_1 c} = \frac{1}{2f},$$

where f is the radio frequency.

For cancellation of the deflection of the unwanted particles

$$\frac{L}{\beta_2 c} = \frac{1}{f},$$

so that $fL = \beta_2 c$ and the condition is achieved for one (or rather a discrete series allowing other multiples of $1/f$) momentum if f and L are fixed. For two contaminants, such as π^+ and p in a K^+ beam, it is still possible to achieve separation at certain momenta by arranging a length such that the two contaminants arrive at R2 at a phase interval of 2π. This fixes the momentum at which the system operates, assuming that there is still adequate deflection of the wanted particles at this momentum.

Once a beam has been set up it is not easy to move components, so that with two separators at a fixed distance apart operation is generally possible only at certain fixed momenta. Complete flexibility is achieved only with three r.f. cavities. The frequency used in the CERN r.f. cavities is 2856 MHz with peak power of up to 20 MW yielding transverse momenta up to ~ 20 MeV/c. This implies deflections of over a milliradian for 16 GeV/c particles, and the system has been used to produce separated K^+ and K^- beams at this momentum.

1.6   Bubble chambers: introduction

For none of the detectors discussed here will we consider more of the technical details than is essential to an understanding of the way in which they can be used in the study of particle physics.

In the bubble chamber an unstable state of liquid is created by superheating. The very small amount of energy deposited by a minimum-ionizing particle (20 MeV m^{-1} or $3{\cdot}2 \times 10^{-12}$ J m^{-1} in liquid hydrogen) is sufficient to trigger off the instability to produce boiling so that the tracks are visible as strings of small bubbles. The superheating of a bubble-chamber liquid is achieved by a sudden reduction of the liquid pressure, either by a system of valves operating on gas in contact with the liquid (gas expansion) or by a piston itself in contact with the liquid (liquid expansion). The properties of the most used bubble-chamber liquids are summarized in Table 1.

The timing of the expansion is arranged so that the beam pulse from the accelerator enters the chamber when the pressure is a minimum and the chamber is at its most sensitive. The flash is delayed for about a millisecond, in order to let the bubbles grow to a size suitable for photography. The cameras are wound on before the next expansion. The cycle is shown schematically in Figure 9. Most

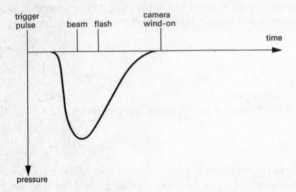

Figure 9 Diagrammatic representation of a bubble-chamber cycle

Figure 10 The 2 m chamber body with the safety cold tank on the camera side; the corresponding cold tank on the flash side cannot be seen in this picture (Photograph by courtesy of Photo CERN)

high-energy proton accelerators produce a pulse only every one or two seconds. It has recently proved possible to kick some particles onto a target serving the bubble-chamber beam more than once during the cycle, while the circulating beam is held at constant energy with a 'flat-top' guide field. The chamber must then be expanded twice at an interval of 300 ms or less, but this has been successfully achieved.

Bubble chambers are invariably used with a high magnetic field orthogonal to the direction of the particle beam and parallel to the optical axis of the cameras. Fields of about two teslas are normal, while a chamber has recently been designed with a field of seven teslas produced by a superconducting magnet.

The largest chambers currently operating have lengths of about two metres with depth and height of about half a metre. Chambers of this size are major engineering enterprises requiring electronic control, and gas and fluid handling of great complexity. The magnets consume megawatts of power, while for the cold chambers a large hydrogen refrigerator is needed. The CERN two-metre chamber is shown in Figure 10 and the British one-and-a-half-metre chamber in Figure 11.

Table 1 Properties of Some Commonly Used Bubble-Chamber Liquids

Liquid	Operating temperature/ K	Operating pressure/kN m^{-2} (p.s.i.)	Density/ g cm^{-3}	Radiation length/m	Index of refraction	Notes
H_2	28	538 (78)	0·06	11·45	1·09	Pure proton target Highly explosive
D_2	32	738 (107)	0·13	9·50	1·1	Simplest neutron target
He	3	3·4 (0·5)	0·14	9·00	1·03	Spin 0, i-spin 0 Hyperfragment source
C_3H_8 (propane)	333	2069 (300)	0·44	1·18	1·22	Highly inflammable
CF_3Br (Freon)	303	1862 (270)	1·5	0·11		Non-inflamable
xenon	252	2551 (370)	2·18	0·035	1·18	Extremely expensive

1.7 Bubble chambers and track analysis

Since the number of photographs taken in a typical experiment is frequently very large (several hundred thousand stereo triplets is regarded as average), an increasing degree of automation is required to cope with their analysis. The photographs are first projected onto a screen in order to note those in which interactions occur. This is still largely carried out manually although machines are now having some success with 'event' finding. The problem of finding events of varying topology among a number of non-interacting tracks is a complicated pattern-recognition task, readily carried out by people but difficult to automate.

Figure 11 The British liquid-hydrogen bubble chamber of 1·5 m installed at CERN in 1964 (Photograph by courtesy of Photo CERN)

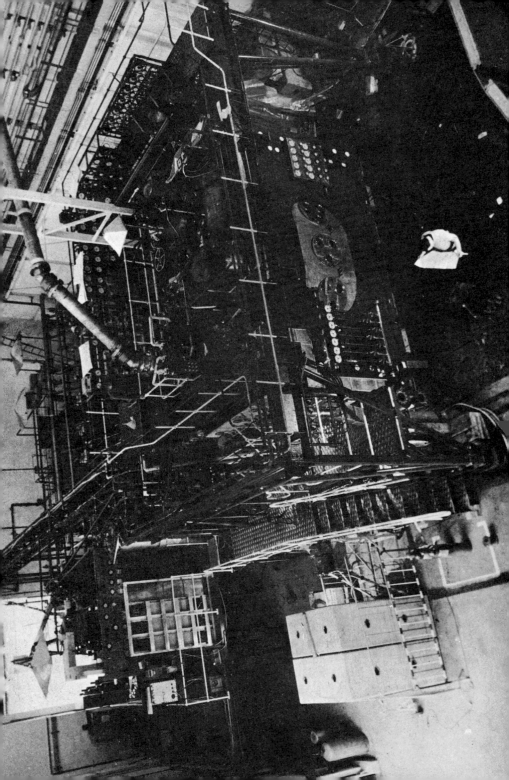

The film is then transferred to a measuring machine on which precise measurements are made of (a) the reference or 'fiducial' marks which are engraved on the chamber windows, or body, and which enable the views from different cameras to be related, (b) the vertices for interactions and decays, and (c) a series of points along each track. These measurements are made separately, on each of the stereo views, to a precision of a few microns on the film. Many different varieties of measuring device have been used for this purpose, ranging from elaborate projection microscopes to automatic optical–mechanical and cathode-ray-tube scanning devices linked to, and controlled by, digital computers. The fastest of these machines are capable of measuring over a hundred stereo triplets per hour.

The coordinates recorded in digital form by these machines require two major stages of routine analysis before they yield the basic data for physics. First the event must be reconstructed in three dimensions. This is achieved by an iterative process in which the distorted helix, corresponding to a track as viewed in the liquid hydrogen through various thick glass windows, is constructed by tracing rays from the images on film. The output from such a 'geometry' program consists of the positions of all vertices and the dip angle, azimuthal angle and curvature of all tracks. From the curvature of the tracks the momentum of the particle may be determined, since the magnetic field distribution is known in detail.

The second stage of analysis consists of the testing of all possible mass assignments to the tracks, consistent with the conservation laws known to hold, such as the conservation of charge, baryon number and strangeness. For each set of assignments or hypotheses, it is then possible to calculate the energy of the particles since the momenta are already known. There are then four scalar constraint equations corresponding to energy and momentum conservation. In practice, if the momenta of all particles are measured, the reaction is so overconstrained (four constraints) that ambiguous hypotheses are uncommon. If a neutral, unobserved particle is present then three equations are necessary to fix the momenta of such a particle, leaving only one equation of constraint. Such hypothesis-testing programs have as their output the best fitted values of momenta for all particles plus a χ^2, or probability, for any successful hypotheses.

The success of this kind of large-scale bubble-chamber physics depends critically on the availability of large digital computers to carry out the geometrical reconstruction and kinematic fitting.

We have discussed the determination of momentum from curvature. Other parameters of the tracks also yield physical information:

Range. The range is a function of particle energy (and charge). For low-energy particles which stop in the bubble-chamber liquid, the range yields the most accurate value of the energy.

Ionization density. In a bubble chamber this is manifest as the number of bubbles, or gaps, per unit length in tracks. In nuclear emulsions the grain density, and in cloud chambers the drop density, provide equivalent information. In each case the ionization varies as $1/\beta^2$ down to a minimum value. At higher momenta in materials of high atomic number the ionization again increases. The 'relativistic increase' has not been detected in a hydrogen bubble chamber. In all detectors the absolute value of the ionization density also depends on other factors. In the bubble chamber it is a sensitive function of the liquid temperature, and bubble-density comparisons are normally made only within the same photograph.

Multiple scattering. The degree of multiple, or small angle, Coulomb scattering of a particle in passing through matter is inversely proportional to the value of $p\beta$ for the particle. The scattering is generally measured as the average deflection over a small path distance. This parameter is not useful in hydrogen bubble chambers, where the multiple scattering is very small, but was commonly used in nuclear emulsions, and is sometimes used in heavy-liquid bubble chambers.

δ-rays. The cross-section for production of δ-rays also varies as $1/\beta^2$.

We may note that in order to determine a particle *mass* we need good data on two of the above characteristics having different dependence on β.

The bubble chamber has many advantages as a particle detector. It provides the fullest information about events, even of high complexity, of any detector. It is sensitive over 4π steradians, the spatial resolution is very good and the precision for the measurement of momenta is fairly high. Its prime disadvantages are that it cannot be triggered to expand only when an event likely to be interesting has been detected by counters, and that its data-acquisition rate is small compared with spark chambers, which can pulse very much faster.

A detailed discussion of all aspects of bubble-chamber principles, construction and analysis is given in Shutt (1967) and Henderson (1970).

1.8   Scintillation counters

A large number of substances have been found to emit light or scintillate when a charged particle passes through them. Scintillators may be in the form of inorganic crystals such as sodium iodide or caesium iodide, both with thallium impurity, organic crystals such as anthracene, organic liquids like toluene, or in the form of plastic (solid solutions).

The light produced in the scintillator, due to ionization, excitation, and possibly also molecular dissociation, arising from the passage of a charged particle, is detected by a photomultiplier. Electrons from the photomultiplier cathode are multiplied by the dynode structure to produce a pulse at the output, which may be amplified and used to drive scalars or to actuate coincidence or other circuits.

Although sodium or caesium iodide crystals are particularly useful for gamma-ray detection, most of the scintillation counters used in high-energy experiments are of organic plastic scintillator material. The plastic may be used in large slabs, or in other shapes, or in the form of a matrix of many thin finger counters, according to the application. Frequently it is coated with reflecting foil to maximize the light passing to the multiplier. It is often inconvenient, or impossible, to place the multiplier close to the actual scintillator, in which case lucite light pipes may be employed

Figure 12 A large plastic scintillation counter with perspex light guides of equal length from all parts of the counter to the multiplier (Photograph by courtesy of Photo CERN)

to link them. By this means, multipliers may be kept away from strong magnetic fields which affect their operation. An elaborate scintillation counter is shown in Figure 12.

Scintillation counters are used in high-energy physics in a multitude of ways. Their high degree of time resolution may be employed to determine particle velocities by the measurement of time of flight between two counters separated by a known distance. Momenta may be measured by detecting coincidences between counters in arrays separated by a magnetic field. The pulse size in a thin counter, in which a particle loses little energy, gives a good measure of dE/dx for the particle. If the particle can be stopped in the counter the pulse size is proportional to E. A combination of E and dE/dx counters can yield the particle mass in regions where dE/dx is not too near minimum.

Scintillation-counter arrangements are particularly useful in experiments on processes which are not too complicated (particularly processes with only two particles in the final state), and where a high rate of data collection is possible. They have the possibility of good time resolution but lack the degree of spatial resolution near a vertex which can be achieved with bubble chambers.

1.9 Spark chambers

Since about 1959 spark chambers have grown steadily in importance as a detection technique in particle physics. There are several different kinds of chamber:

1.9.1 *Narrow-gap or track-sampling chambers*

Narrow-gap chambers consist of an arrangement of accurately parallel plates, or foils, separated by gaps of about 0·2–1 cm and usually filled with a noble gas, such as neon at a pressure of about one atmosphere. A high-voltage pulse, to produce a field of about 10^6 V m^{-1}, is applied to alternate plates when a suitable arrangement of scintillation counters indicates that a track has passed through the chamber, and that an event of the required variety has occurred. The ionization in the gaps is sufficient to trigger a spark discharge between the plates. A clearing field is used to remove the ions before another particle can be recorded.

The sparks are self luminous so that their positions may be readily recorded photographically. Various arrangements which may be used to see between the plates are shown in Figure 13. Orthogonal viewing is possible, and mirrors are frequently used to photograph both views on the same frame. An example of tracks photographed in a parallel-plate chamber in a magnetic field is shown in Figure 14, while a typical large spark-chamber array is shown in Figure 15. Other methods of recording the spark positions have been developed which avoid the need for the photographic film as an intermediate medium between the chamber and the digital measurement. One such method is the use of two or more microphones which locate the spark by sound ranging. The microphones are generally in the form of piezoelectric crystals or capacity transducers. When the counters signal

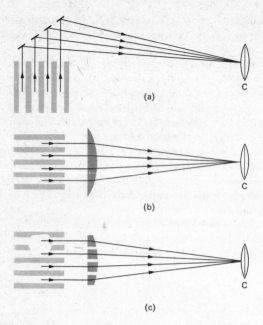

Figure 13 Optical arrangements used for spark chambers. C is the camera objective. In (a) a system of strip mirrors reflects the light into the lens. In (b) a large lens is used. In (c) the lens is replaced by a series of lens sections

the passage of a particle through the chamber, a digital clock is started. The time of arrival of the sonic pulse at each transducer is a measure of the distance of the spark from that transducer. The use of more than two transducers is necessary to discriminate against the presence of more than one spark. The delays may be recorded by a computer, or the actual position computed immediately and recorded on magnetic tape.

An alternative filmless technique is the *wire chamber*. In this device the face of the chamber consists of an array of closely spaced fine parallel wires. The wires may

Figure 14 A K^0 produced in $K^-p \rightarrow K^0n$ in a magnetic spark chamber (Astbury *et al.*, 1964)

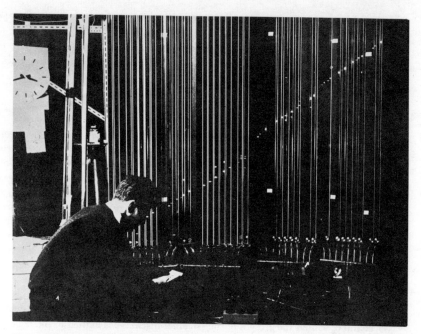

Figure 15 A cosmic-ray track passes through a large spark-chamber array at CERN (Photograph by courtesy of Photo CERN)

then be each threaded through a magnetic core. A spark passing from a wire flips the core so that the spark-position information is immediately recorded, and can then be transferred to the computer. Magnetostrictive read-out for wire chambers is also used. A magnetostrictive wire or ribbon (nickel) is magnetized to a bias level such that it is magnetically coupled to the wires. The magnetic pulse associated with the spark causes a local elongation of the ribbon which propagates down it with the speed of sound (about 5 mm s^{-1}). The wire associated with the spark may then be easily identified since a sensitivity of 0·02 μs can be achieved.

Wire chambers have certain advantages in addition to the direct transfer of information to the computer. Less energy is required for this technique than is needed for a full-blown visible spark so that a faster recovery time is possible (a few hundred microseconds). Also, in the parallel-plate chambers it is difficult to achieve uniform sensitivity when more than one spark discharges; this 'robbing' problem is much less severe in wire chambers. Wires are usually ~ 0.05–0.25 mm in diameter with spacing around 1 mm.

1.9.2 Wide-gap chambers

Wide-gap chambers may have a distance of 0·3–0·4 m between the plates and have been used for tracks at angles up to 45° with the normal to the electrodes, although beyond about 20° the spark intensity falls off. Fields of 400 kV m^{-1} are normal, so that for large chambers potentials of 100–200 kV are needed. Pulsers (known as Marx generators) for such fields generally operate by charging a bank of condensers in parallel, and discharging them in series by triggered spark gaps. These chambers can also operate in a magnetic field so that particle curvatures can be measured.

1.9.3 *Projection and streamer chambers*

In the chambers we have discussed above, the sparks are viewed in directions normal to the plane of the electrodes, and the particles must pass within a *maximum* of about 45° to the normal to the electrodes. In a *projection chamber* one of the electrodes is made transparent, in the form of a wire screen or conducting glass plate. The chamber is viewed through this electrode, parallel to the field. Tracks pass roughly parallel to the plates, and every spark passes across the gap, so that viewed parallel to the plates a track appears as a sheet of discharge. The chambers can be used with a number of simultaneous tracks, and have a short recovery time so that rates of 10^4 pulses per second are possible. The pulse energy is generally small and the light emission low, which limits the depth of field since large lens apertures are needed.

The *streamer chamber* is the nearest approach to a truly isotropic spark chamber. This has been achieved by applying a very short high-voltage pulse to a wide-gap chamber so that the growth of the streamer discharge is limited to the region near its origin. The streamer length is a function of the pulse length, and a pulse of about 50 ns is needed to achieve streamers of a few millimetres in a 10^4 kV m^{-1} field. Such chambers may be viewed, like a bubble chamber, by a pair of stereo cameras at angles near to the field direction.

Compared with bubble chambers, spark chambers have the advantage that they can be selectively triggered, by an arrangement of counters, so that the events recorded are very largely of the kind wanted for a given study. The spark chamber has also a high data-collection rate with the possibility of ten pictures during an accelerator pulse lasting 400 ms. The triggering property means that the beams necessary for spark-chamber experiments can often rely on electronic mass separation by time-of-flight or Cerenkov counters, rather than elaborate electrostatic or r.f. techniques. Spark chambers are also the cheapest detectors, per unit volume, with good spatial resolution.

Compared with bubble chambers, however, spark chambers do not have the advantage of identical target and detection medium, particularly important for short-lived particles, nor the isotropic sensitivity.

A more detailed discussion of spark chambers and some of their applications can be found in Cronin (1967) which also gives detailed reference to earlier work.

1.10 Cerenkov counters

If a particle passes through a medium in which its velocity is greater than that of light, in the medium, it will emit electromagnetic radiation. This phenomenon was first observed by Cerenkov, and bears his name. Cerenkov counters use a photo-multiplier to detect the radiation in the visible region, and thus to record the passage of a particle. Compared with scintillation counters Cerenkov counters are less sensitive, and more clumsy, but they possess the property of discrimination between particles of different velocity.

Basically Cerenkov counters consist of a container filled with a gas or liquid, or a solid material which is transparent to the Cerenkov light, and for which the index of refraction μ is suitably chosen. For the emission of Cerenkov light we require

$$\mu > \frac{c}{v}$$

where v is the velocity of the particle to be detected. This simple condition under-lies the operation of the *threshold Cerenkov detector*. Suppose that we wish to dis-tinguish K^+ mesons at 1 GeV/c from protons and π^+ mesons. β for protons, kaons and pions at this momentum is 0·73, 0·89 and 0·99 respectively. The corresponding values for μ are 1·37, 1·12 and 1·01. If we then arranged two Cerenkov counters, one filled with water and one with carbon dioxide at a pressure and temperature such that $\mu_{CO_2} = 1·05$, the protons would record in neither, the kaons only in the water counter, and the pions in both.

The Cerenkov radiation is emitted along the surface of a cone the apex of which is the instantaneous position of the moving particle, and the angle of which is given by

$$\theta = \cos^{-1}\left[\frac{1}{\beta\,\mu(v)}\right], \qquad\qquad\qquad \textbf{1.2}$$

where $\mu(v)$ is the index of refraction of the material for light of frequency v. The emitted light is plane polarized such that the electric vector $\mathbf{E}$ is in the plane of the incident particle and the emitted photon. For a single charged particle I, the number of photons emitted per second per unit path length per unit frequency, is given by

$$\frac{d^2 I}{dx\,dv} = \frac{4\pi^2 e^2}{hc^2}\left(1 - \frac{1}{\beta^2\,\mu^2(v)}\right) = \frac{4\pi^2 e^2}{hc^2}\sin^2\theta \qquad\qquad \textbf{1.3}$$

$$= \frac{2\pi}{137c}\sin^2\theta.$$

If this formula is evaluated for water, say, it is found that the light emitted per centimetre is only $\frac{1}{44}$ of that from a scintillation counter.

The condition **1.2** is exploited in the focusing Cerenkov counter where there is discrimination on the angle of emission of the Cerenkov light. Figure 16 shows diagrammatically the arrangement in a focusing counter. Light emitted at the angle corresponding to the desired velocity is reflected by the walls of the radiator until it escapes. A cylindrical mirror and baffle allow only light from a limited region of emission angle to reach the photomultiplier. With such a counter it is possible to select the required β from higher and lower values.

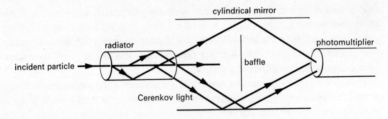

Figure 16

For good resolution in such a counter it is desirable to minimize $\partial\theta/\partial\beta$. Equation **1.2** yields

$$\frac{\partial\theta}{\partial\beta} = \frac{1}{\beta^2\mu\sin\theta}.$$

Thus best resolution is achieved with small θ. However, **1.3** shows that the intensity is proportional to $\sin^2\theta$. Practical counters for any given experimental situation must make the best compromise between these factors.

For values of β very near to unity, gas-filled Cerenkov counters are often used. These have the advantage that μ can be varied by simply changing the gas pressure. With such counters kaons can be clearly distinguished from pions even at 8 GeV/c.

Since electrons become relativistic at very low momenta, Cerenkov counters are often used as electron detectors, especially in the presence of a high, non-relativistic background. Gamma rays are also frequently detected by allowing them to produce showers in lead glass, the shower electrons giving rise to the Cerenkov light. With suitable design the emitted light can give a measure of the gamma-ray energy.

Further details concerning Cerenkov counters may be found in Lindenbaum and Yuan (1961).

1.11 Other detectors

We have not discussed a number of other detectors which find only occasional use, at the present time, in elementary-particle physics.

The nuclear emulsion was one of the most important tools in the early days of particle physics. These high silver concentration photographic plates can record the tracks of even minimum-ionizing charged particles. They are small and continuously sensitive, and were ideal for cosmic-ray work particularly with high flying balloons. The position measurement precision in emulsions is very high, so that if particles stop in the emulsion the range gives a precise measurement of energy. Grain-density measurements can yield velocities for non-relativistic tracks, while multiple scattering was widely used to yield values of $p\beta$ which could be combined with the grain density to determine the mass. However, bubble chambers operating in high magnetic fields, and containing pure hydrogen or deuterium in contrast to the complex constitution of the emulsion (Ag, Br, H, N, O), have superseded emulsions in most applications. Emulsions are still used for studies of hyperfragments.

Cloud chambers were one of the most important tools in nuclear physics and also, in the early days, of particle physics using cosmic rays. They retain the advantage over bubble chambers that they can be triggered by counters. The fact that the low density of the gas allows very precise energy measurement for low-energy particles is of little value in high-energy physics. In all other respects in high-energy work the bubble chamber is superior.

Solid-state counters have found little application at high energies, but the new wire proportional counters of Charpak, if they can be manufactured on a large scale, may prove superior to spark chambers for many purposes.

Chapter 2
π-Mesons and μ-mesons

2.1 Prediction of the π-mesons by Yukawa

The prediction of the existence of mesons, by Yukawa in 1935, and their subsequent discovery in the cosmic radiation, present one of the most striking examples of the interaction of theory and experiment in modern physics.

We can understand Yukawa's argument in a qualitative way as follows. It had become well established that electromagnetic forces could be well understood in terms of a field of which the *quanta* were photons. In this theory of quantum electrodynamics the forces between two charged particles are attributed to the exchange of photons between them. Thus, for instance, in the centre-of-mass system (see Appendix A.2) the Coulomb scattering of two electrons is described by a diagram like

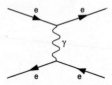

Figure 17 (a)

where energy and momentum have been exchanged by the exchange of a photon. This diagram is similar to a so-called Feynman diagram, which we shall have cause to use quite frequently. The more usual convention is to represent the process simply as

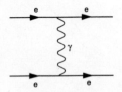

Figure 17 (b)

Let us first note the broad qualitative features of nuclear and electromagnetic forces.

Evidence is available from *static nuclear properties* and from *scattering experiments*. Our purpose here is not to treat these subjects in detail, but rather to discuss briefly those aspects which are relevant to the size of the nucleus and the strength of the nuclear forces.

It is immediately clear that the nuclear forces are attractive and sufficiently strong to overcome the Coulomb repulsion. It is also clear, since the nuclear radius R obeys the relation

$$R = r_0 A^{\frac{1}{3}},$$

(where r_0 is a constant and A is the mass number) that there is some form of 'repulsive core' which prevents the collapse of all the nucleons to the range of the nuclear force. We note also that the binding energy of nucleons in nuclei is about eight million electronvolts throughout the greater part of the periodic table, i.e. the nuclear force is 'saturated'. This behaviour is similar to that observed for the inter-molecular forces in liquids and solids. It is in contrast with the unsaturated Coulomb force, responsible for the binding of electrons in atoms, which are all of about the same size throughout the periodic table. The saturation is characteristic of so-called *exchange forces*. We may describe such a force in terms not of a simple potential $V(r)$ but rather of the product $P_{12} V(r)$, where P is the so-called permutation operator. Thus, if the wave function of the pair of interacting particles is $\psi(1, 2)$, then the effect of P_{12} is to exchange certain of the properties of the particles. Thus

$$P_{12} \psi(1, 2) = \psi(2, 1).$$

The exchange may be one of charge, spin or position, or of any combination of these properties. The suggestion that the inter-nucleon force was due to some kind of exchange was first put forward by Heisenberg as early as 1932, although the nature of the exchange 'quantum' was not clear.

Before discussing further the magnitude of the range and strength of the forces, we may look at what evidence there is for their exchange nature from scattering experiments. Let us consider the scattering of neutrons on protons since in this case the complication of the Coulomb force is absent, as is the problem associated with having two identical particles. If we consider the scattering of neutrons of moderately high energy (say 100 MeV), then in the case of no exchange force the scattered neutrons will be peaked in the forward direction, in the centre-of-mass system, since we will have contributions to the scattering from states having orbital angular momentum greater than zero. For exchange forces, on the other hand, the identity of the particles will be exchanged in the interaction and we will have protons peaked in the forward direction. In fact the experimental results show a peak at both 0° and 180°, indicating the existence of both exchange and non-exchange forces.

An important question is whether the forces between the three possible pairings of nucleons, n–n, p–p, and n–p are the same, apart from the Coulomb forces between the protons. Information on this point can be obtained from a comparison of the masses and energy levels of so-called mirror nuclei. Such pairs of nuclei differ only in that a neutron in one is replaced by a proton in the other. Thus, if the n–n and p–p forces are the same, with the exception of the Coulomb force, the masses and energy levels of, for instance, $^7_3$Li and $^7_4$Be or $^{11}_5$B and $^{11}_6$C should be the same when a correction has been made to take into account (a) the Coulomb effects and (b) the neutron–proton mass difference. Such a comparison in fact confirms the equality of the n–n and p–p forces. This property of nuclear forces is known as *charge symmetry*. If these forces are also equal to the n–p force, then we have *charge independence*. This is more difficult to establish from the static nuclear properties, but is supported by a comparison of isobars with even A in which the charge differs by two units. We shall return to this subject when we come to consider the property known as isotopic spin.

We now consider some quantitative evidence as to the range and strength of the nuclear force.

Ranges from nuclear size. We can get some indication of the range of the nuclear force if we recall that many experiments have established that the nuclear radius follows closely the relation

$$R = r_0 A^{\frac{1}{3}},$$

where $r_0 = 1 \cdot 5$ fm.

If we imagine that in nuclei the inter-nucleon distance is of the order of the range of the nuclear force, and that the nucleons are uniformly distributed so that each nucleon occupies a volume equal to πr_0^3, then the range of the force is also of the order of r_0. Clearly such a consideration involves several crude approximations and assumptions and only gives an order of magnitude value.

Alpha-particle scattering on nuclei. The original results on the scattering of alpha particles by nuclei, which established the nuclear atom, were well fitted on the assumption that the scattering was due to the Coulomb force between the alpha particle and the nucleus, considered as point charges. However it is clear that the effects, both of the nuclear size and of the nuclear forces, will be important if the closest distance of approach becomes sufficiently small. This is indeed found to be true. For scattering of alpha particles by gold, for instance, the cross-section falls dramatically below the value to be expected for Coulomb scattering for an energy greater than about 20 MeV. Various techniques have been used to attempt to fit the data in terms of the nuclear radius and the range of the nuclear force as parameters. Again the value for the range of the force is found to be of the order of $1 \cdot 5$ fm.

Neutron–proton scattering. The study of this phenomenon would appear to be particularly suitable for obtaining the range since Coulomb effects are absent and we are dealing with a two-body situation. The simplest approach is to assume that the scattering can be described in terms of a square-well potential. In fact changes in the potential shape have only small effects on the results. The parameters to be determined are then the potential depth V_0 and the well diameter d, where we consider neutron energies such that only s-state scattering is possible (less than ~ 5 MeV). Although we know the binding energy of the deuteron, and can measure the neutron–proton scattering cross-section, we cannot determine V_0 and d separately, because the scattering takes place in both the singlet and triplet spin states, so that in fact there are four parameters rather than two: $V_{0,t}$, d_t, $V_{0,s}$ and d_s. The result of fitting the data in terms of these four parameters is

$V_{0,t} \simeq 35$ MeV $\qquad d_t \simeq 2$ fm

$V_{0,s} \simeq 12$ MeV $\qquad d_s \simeq 3$ fm.

Low-energy proton–proton scattering also yields values

$V_{0,s} \simeq 13$ MeV $\qquad d_s \simeq 2 \cdot 6$ fm.

Deuteron binding energy. The application of the Schrödinger equation to the case of the deuteron, which, on the basis of the value of the magnetic moment, can be taken to be largely in the triplet s-state, yields a relationship between the potential and the range of the form

$$V_0 d^2 \simeq \frac{\pi^2 \hbar^2}{4M},$$

where M is the nucleon mass. Taking $V_0 \simeq 35$ MeV we obtain $d \simeq 1 \cdot 8$ fm.

Summarizing, we can say that any satisfactory theory of nuclear forces had to account for a force which was very strong (well depth ~ 35 MeV), of short range (~ 2 fm) and which acted equally between proton–proton, neutron–neutron and proton–neutron. The force had also the characteristics, at least in part, of an exchange interaction, and the success of such a theory for electromagnetic forces made its pursuit particularly attractive also for the nuclear interaction.

Yukawa developed this idea in his famous paper of 1935. He first noted that a zero-mass quantum, such as the photon, will not give a force of sufficiently short range. In addition, since the quanta of the field must certainly be permitted to travel at high velocity, it is essential that the theory be relativistically correct. We can illustrate the results of the Yukawa treatment in the following way.

The relativistic relationship between energy, momentum and mass for the field quantum of mass m is

$E^2 - p^2 c^2 - m^2 c^4 = 0.$

We can now use the usual quantum-mechanical substitutions

$E \rightarrow i\hbar \dfrac{\partial}{\partial t} \quad$ and $\quad p \rightarrow i\hbar \, \mathbf{\nabla}$

to produce an operator equation

$-\hbar^2 \dfrac{\partial^2}{\partial t^2} + \hbar^2 \, \mathbf{\nabla}^2 c^2 - m^2 c^4 = 0.$

If we now represent the force between nucleons by a potential $\phi(r, t)$, which may be regarded as a field variable, then we have

$$\left[\nabla^2 - \frac{1}{c^2}\frac{\partial^2}{\partial t^2} - \frac{m^2c^2}{\hbar^2}\right]\phi = 0$$

as our 'wave equation'.

The time-independent part of the equation is

$$\left[\nabla^2 - \frac{m^2c^2}{\hbar^2}\right]\phi = 0, \tag{2.1}$$

to be compared with the equation for a static electric field,

$$\nabla^2\phi = 0,$$

with solution $\quad \phi = \dfrac{e}{r}$.

The solution of the differential equation **2.1** is

$$\phi = \frac{g}{r}e^{-mcr/\hbar},$$

as can be checked by substitution, where g is a constant with the dimensions of electric charge, and is known as the *coupling constant* for the interaction. Thus the energy of a second nucleon in this field ϕ generated by the first is $g\phi$ and the interaction between the two nucleons is given in terms of the coupling constant by

$$\frac{g^2}{r}e^{-mcr/\hbar}.$$

If we write the form of the potential as proportional to e^{-ar}/r we see that to a

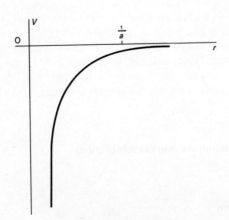

Figure 18 The form of the Yukawa potential

first approximation the force becomes zero at $r = r_0 = 1/a$. Thus the range of the force, r_0, is $\sim h/mc$. If we substitute 2 fm for our r_0 we find $m \simeq 200 \text{ MeV}/c^2$ for the mass of the field quantum. We note that as $r \to 0$, $\phi \to \infty$ and the form of this 'Yukawa potential' is shown in Figure 18. Although this is certainly not a square well we can use the value of V_0 obtained from the consideration above to obtain a value for the coupling constant. This yields $g^2 \simeq \hbar c$. It is clear that the potential cannot be infinitely deep so that the form which we have derived cannot apply at very small values of r. However, we may hope that it is at least approximately true at the ranges with which we shall be initially concerned.

If we consider the various inter-nucleon forces in terms of the exchange of the Yukawa quanta, which we will refer to as π-mesons or pions, in anticipation of their later discovery, we obtain the following results. For proton–proton and neutron–neutron scattering exchange of a neutral pion is required, unless exchange of two-charged mesons is allowed.

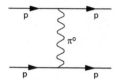

Figure 19 (a)

For neutron–proton scattering, however, we may have exchange of both neutral and charged pions.

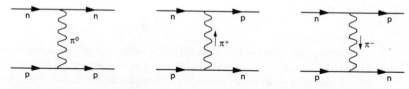

Figure 19 (b), (c), (d)

The equality of the n–n, n–p and p–p forces indicates that all are due to the same type of exchange, so that we must suppose that neutral, as well as charged, mesons should exist. This extension of the original Yukawa proposal was made in 1938 by Kemmer.

Finally, in predicting the properties of the pion Yukawa pointed out that it was probably unstable. This proposal arose from attempts to explain nuclear beta decay, and to account for the fact that the mesons had not so far been observed. His explanation of the beta decay, in terms of the steps

$$n \to p + (\text{meson})^- \quad \text{and} \quad (\text{meson})^- \to e^- + \nu,$$

is now known to be wrong, though in fact the meson lifetime he obtained on this hypothesis was by chance approximately correct. There is apparently no selection rule to prevent the pion from decaying via electrons, neutrinos and gamma rays.

The picture of the nucleon which is implied by the Yukawa model is of a particle continually emitting and reabsorbing pions, so that it is effectively surrounded by a pion cloud. A single nucleon cannot emit a pion with conservation of energy, except within the limits allowed by the uncertainty principle. Thus a violation of energy conservation by an amount ΔE can exist only for a time Δt given by $\Delta t \, \Delta E \simeq \hbar$. If $\Delta E = M_\pi \, (\sim 140 \text{ MeV})$ then $\Delta t \simeq 4 \times 10^{-24}$ s. For velocity c this gives the extent of the pion cloud as $\sim 10^{-12}$ mm.

A further interesting application of this picture of the meson–nucleon relationship is to the problem of the anomalous magnetic moment. Dirac showed that it was a consequence of his relativistic wave equation that the electron should have a magnetic moment given by

$$\mu_e = \frac{-e\hbar}{2m_e c}.$$

The Dirac theory would then give

$$\mu_p = \frac{e\hbar}{2m_p c}$$

and $\mu_n = 0,$

for a proton and a neutron respectively.

In fact we have,

$$\mu_p = 2 \cdot 78 \mu_N \quad \text{and} \quad \mu_n = -1 \cdot 93 \mu_N,$$

where $\mu_N = \dfrac{e\hbar}{2m_p c},$

the nuclear magneton.

The differences between the predicted Dirac values and the measured values are known as the anomalous magnetic moments. The meson picture affords a qualitative explanation of the observed magnetic moments. Since we assume that for part of the time the proton exists as a neutron plus a π^+ meson,

$$p \rightleftarrows n + \pi^+,$$

we must consider the effective magnetic moment for the $n\pi^+$ system. We will assume that the pion has zero spin (see section 2.6) so that if the relative orbital angular momentum of neutron and π^+ is $L\hbar$, then

$$\mu = \frac{e\hbar L}{2m_\pi c},$$

and for the p-state this gives $\mu = 7\mu_N$. The true μ_p is then given by

$$\mu_p = x\mu_N + 7(1-x)\mu_N,$$

where x is the probability that the proton exists in the so-called bare state, with no external pion. Using the measured value for μ_p we find $x = 0.7$. Considering the inverse process, where the neutron exists for part of the time as a proton plus a π^- meson, and again taking the pion spin to be zero, we obtain the effective magnetic moment for the p–π^- combination by adding algebraically the true bare proton moment and that generated by the 'orbiting' pion. If we assume that x is here also equal to 0.7, and note that the orbital moment and the bare proton moment must be in opposite directions for the p-state in order to obtain spin $\frac{1}{2}$, we find that $\mu_n = -2.4\mu_N$ in qualitative agreement with the observed value.

In fact, of course, the hypothesis that nuclear forces are due to the exchange of π-mesons does not explain all their features, such as for instance the spin dependence. We now know that there exist a number of other strongly interacting mesons which must certainly also play a similar role to the pion in nuclear forces. However the pion is the lightest of such mesons and must therefore account for the most important contribution at the outer edge of the potential. At the moment the fact that the situation has turned out to be much more complicated than envisaged by Yukawa is not important to us; the basic idea remains valid, and it was this idea which both stimulated the later experiments and enabled them to be interpreted in terms of the quantum of the nuclear force.

2.2   Discovery of π-mesons and μ-mesons

In the late 1930s, following the suggestion of Yukawa, the only source of particles of sufficiently high energy to produce the mesons was the cosmic radiation. Evidence for the existence of mesons was forthcoming from both counter studies of the absorption of the soft and hard components of the cosmic rays and, even more strikingly, from cloud-chamber photographs.

In discussing the cloud-chamber evidence we first consider the possible information concerning a particle mass which can be obtained from measurements on a track in a cloud chamber. Measurements may be made of ρ the radius of curvature in a magnetic field, of R, the range, if the particle comes to rest in the chamber, of the drop density and of the change in curvature of the track in passing through absorbing material, normally in the form of a plate across the chamber. We recall that ρ is a function of the particle momentum, R is a function of the particle energy, and the drop density or rate of energy loss, of the particle velocity. Thus a measurement of ρ and R, or of either of these quantities along with drop density or change in curvature on passing through a plate, will, in principle, yield a value of the particle mass. In practice, the precision obtained may be highly dependent on the energy of the particle and also on such experimental factors as possible turbulence in the cloud chamber, or non-uniform sensitivity.

The first clear evidence for the existence of a particle with mass intermediate between the electron and the proton was obtained by Anderson and Neddermeyer (1936), in cloud-chamber studies of cosmic rays at mountain altitude. In Figure 20 is shown one of the two photographs which they obtained showing the tracks of such a particle. A number of particles are ejected, from the same point in a lead plate

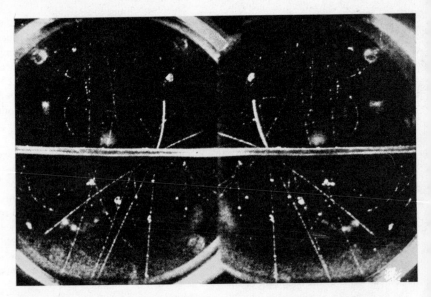

Figure 20 A stereoscopic pair of cloud-chamber photographs taken by Anderson and Neddermeyer. A neutral cosmic-ray particle causes an interaction in the lead plate across the chamber. A densely ionizing track emerging in the upper half of the photograph has a range of 40 mm and a radius of curvature of 65 mm. These data indicate a mass ~ 180 MeV/c^2

across the chamber, by an incident neutral particle. One of the tracks is densely ionizing and stops in the chamber after a range of 40 mm. Its radius of curvature in the 632 kA m^{-1} field is 65 mm. A proton having this range would have an energy of 1·5 MeV and a radius of curvature of 0·2 m. As can be seen from some of the small spirals in the chamber, electrons having a range of this order have much smaller radii of curvature and are minimum-ionizing tracks. In fact if the measured range and curvature are used the mass of the particle is found to be about 180 MeV/c^2. A number of other such observations demonstrated the existence of such particles, having both positive and negative charges. The errors in the masses obtained showed wide variation, but indicated an average mass of about 100 MeV/c^2.

Concerning the meson decay a cloud-chamber photograph was obtained in 1940, by Williams and Roberts, which showed a positive particle of mass about 120 MeV/c^2 decaying into an electron. The evidence concerning the lifetime of the meson came, in the early work, from measurements with Geiger counters on the absorption of cosmic rays. Indeed these measurements also provided independent, indirect evidence on the actual existence of the particles. It was well established that the cosmic radiation consisted of two components known as the hard, or highly penetrating, component and the soft component, which was absorbed almost completely by 0·10–0·15 m of lead. The soft component was responsible for most of what were known as cosmic-ray 'showers', and was observed to consist of electrons, positrons and gamma rays. The hard component was observed to be absorbed roughly in proportion to the mass of the absorber, unlike the soft component in which absorption per atom was approximately proportional to the square of the atomic number. On the assumption that the hard component consists of very-high-energy electrons and positrons, it is possible to calculate the energy which these particles must have if they are to penetrate through the complete atmospheric layer. Such a calculation yields values of the energy which appear to be impossibly high, and for which there is no evidence in the energy spectrum of electrons and positrons observed at accessible altitudes. Initially it was thought that the explanation of this difficulty must lie in a breakdown of the energy-loss formula for particles of very high energy. However, as the formula was tested experimentally at increasing energies, it became clear that there is no evidence for such a breakdown. On the other hand, not more than a very small fraction of the particles making up the hard component can be protons since all attempts to detect protons in the cosmic rays agreed that, at least near the earth's surface, the proportion of protons in the penetrating component is not more than about 10 per cent. Thus the most plausible explanation of the nature of the hard component is that it is due to particles having a mass much greater than electrons, and for which the bremsstrahlung energy loss is thus very much less. ,

Detailed studies of the absorption of the hard component indicated that the absorption was not strictly proportional to the mass of the absorber, but also depended upon its thickness. Experiments of this nature used Geiger counters surrounded by sufficient lead to eliminate the soft component. In such an experiment the absorption of the hard component in air was measured by making counts at different altitudes. A mass of some other absorber, such as carbon, equivalent to the air layer between the two altitudes could then be inserted above the counters. It was found that the air apparently absorbed more effectively than the carbon. This effect finds a natural explanation if we assume that the particles undergo radioactive decay, and quantitative comparison of the rates of absorption yielded a mean lifetime of approximately two microseconds.

A direct determination of the meson lifetime was made by Rasetti. The mesons were selected by a fourfold system of Geiger counters and allowed to pass into a ten-centimetre-thick iron absorber. Mesons which stopped in the iron, as indicated by an anticoincidence of a further set of counters, were found in about half of all cases to be associated with a charged particle emerging from the iron block, after a delay of the order of a few microseconds or less. If this delayed particle is taken to be the decay electron then the lifetime of the meson may be obtained from the distribution in delay times, and again a figure of about two microseconds was obtained. The fact that only half the incident mesons were observed to decay may be explained by assuming that negative and positive particles are present in equal numbers, that the positive particles suffer Coulomb repulsion from the nucleus, so that they are free to decay, but that the negative particles are rapidly absorbed.

Initially these mesons were taken to be the particles predicted by Yukawa since they corresponded roughly, in both mass and lifetime, to his predictions. However increasing doubts arose concerning the interaction cross-section of these particles with nuclei. For the meson to play the appropriate role in nucleon forces it is essential that it has 'strong interaction' (see Chapter 4) with nucleons. Even in the early experiments a difficulty was apparent in the assumption that mesons could pass through great distances in the atmosphere without interaction. If we assume that their cross-section for interaction is of the usual order of magnitude for 'strong' interactions ($\sim 10^{-24}$ mm^2) then we expect to find very few mesons remaining at ground level. More direct evidence was obtained, in 1947, in an experiment by Conversi, Pancini and Piccioni, who studied the absorption of negative mesons brought to rest in carbon and in iron. Broadly speaking, in carbon all the mesons were observed to decay, while in iron they were all absorbed without decay. Further observation proved that the capture rate for these mesons at rest was proportional to the atomic number, Z. At $Z = 12$ approximately half of the mesons decayed while half were captured, indicating an average time for capture of about one microsecond. Such a long capture time may be seen to be in strong disagreement with the hypothesis that the meson–nucleon interaction is 'strong'. The process of capture for the negative mesons is that the slow meson falls into an electron orbit and rapidly cascades downwards into the lowest state. We note that due to its greater mass the Bohr orbit for the meson is about two hundred times smaller than the corresponding orbit for an electron. At $Z = 12$ we can calculate that the meson will spend about 10^{-3} of its time actually in the nucleus. If we assume an interaction cross-section per nucleon of about ten millibarns, and a velocity of the order of c for the meson, then we obtain by crude classical arithmetic a reaction time of about 10^{-23} s for a strong interaction, differing by a factor of 10^{17} from the observed value and indicating that the absorption is a 'weak' process for which the time would be expected to have about the observed value.

These difficulties led Bethe and Marshak, in 1947, to suggest that there must exist another meson corresponding to the Yukawa particle. Conclusive evidence in support of this suggestion was obtained in the same year by Lattes, Muirhead, Ochialini and Powell, who observed the actual decay of one meson into another in a nuclear emulsion. Subsequent work revealed many examples of events like that shown in Figure 21, in emulsions flown for periods at very high altitudes by balloon. The masses of the particles in the three-particle decay chain of the figure were found to be ~ 140 MeV/c^2, and ~ 100 MeV/c^2, with the final particle identifiable as an electron.

We note here the difference in the measurements used for mass determination in a nuclear emulsion and in a cloud chamber. In the emulsion it has not normally been practicable to use a magnetic field to determine particle momentum. We may still obtain the energy from the range, and the velocity from the ionization, of which the grain density in the track is a function. In the emulsion, a useful parameter is the multiple scattering of the track due to Coulomb scattering on emulsion nuclei. This may be measured by evaluating the deviation of the track from a straight line for a series of segments along its length. The mean value of such deviations is a function of the product of momentum and velocity.

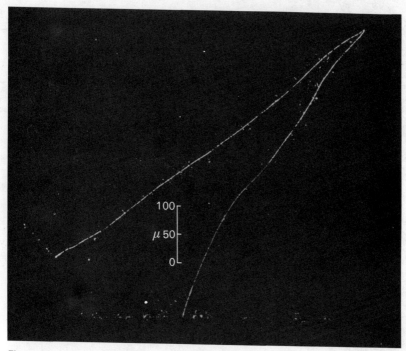

Figure 21 A $\pi \to \mu \to e$ decay in nuclear emulsion. The pion enters the picture at bottom centre, slows down and stops at top right. The muon travels to bottom left before decaying to an electron. The scale is in microns

The parent meson now known as the π-meson, or pion, appears to fill the role of the Yukawa particle, and is not normally observed in cosmic rays at ground level. It is observed to decay into the lighter μ-meson, which forms the major part of the hard component of the cosmic radiation, and which in turn decays into an electron of the appropriate charge. Each of these decay processes must be accompanied by one or more neutral particles which will be discussed below.

2.3 Properties of charged π-mesons and of μ-mesons

2.3.1 *Meson production*

As we have seen, certain properties were determined for the π- and μ-mesons, at least approximately, in the early cosmic-ray work. But the most accurate determinations have depended on experiments using mesons produced by high-energy accelerators. The Yukawa picture would suggest that it should be possible to produce pions in nucleon–nucleon collisions if the bombarding energy is high enough. We may picture the incident nucleon interacting with a pion in the 'cloud' and actually knocking it free.

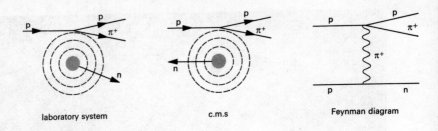

laboratory system   c.m.s   Feynman diagram

Figure 22

The threshold for this process is most easily worked out as follows (see also Appendix A). We use the invariance of the quantity

$$E^2 - p^2 = m^2,$$

where E, p and m are the total energy, momentum and mass of a particle, or system of particles, and where we have taken $c = 1$. If we indicate centre-of-mass (c.m.s.) quantities by dashed symbols, then, since in the c.m.s. $p' = 0$, we have

$$E'^2 - p'^2 = E'^2.$$

In the c.m.s. the threshold for

$$p + p \rightarrow pn\pi^+,$$

or $\quad p + n \rightarrow pp\pi^-,$

or similar processes occurs when

$$E' = 2m_n + m_\pi,$$

where m_n and m_π are the nucleon and pion masses respectively. Also if we denote the incident proton kinetic energy by the symbol T then we have for the incident proton

$$2m_n T + T^2 = p^2.$$

This gives for the minimum kinetic energy necessary to produce a π-meson,

$$T = m_\pi \left[2 + \frac{m_\pi}{2m_n} \right]$$

$$= 285 \text{ MeV.}$$

Thus a cyclotron capable of accelerating to energies of the order of 300 MeV or greater should produce π-mesons when the protons strike stationary nucleons. In fact the threshold for meson production for protons striking a cyclotron target is only about 180 MeV due to the great increase in the c.m.s. energy when head-on collisons occur, in the nucleus, between an incident proton and a moving nucleon.

2.3.2 Pion mass

A good determination of the pion mass has been made by a magnetic analysis of the particles coming from the cyclotron target, to give precise values of their momentum, following which the particles are brought to rest in nuclear emulsions (see Figure 23). The energy can then be very precisely established from the range,

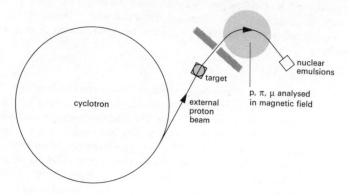

Figure 23 Particles emitted from an external target are momentum-analysed in a magnetic field. The range is then measured in nuclear emulsions to give a precise value of the mass

and a comparison made between protons, pions and muons. The results of this experiment yielded mass values of 139·6 MeV/c^2 for positive and negative pions and 105·6 MeV/c^2 for positive muons.

A particularly accurate value for the mass of both pions and muons may be obtained from measurements of mesic X-rays. We have already discussed the absorption process for slow negative mesons, which proceeds via absorption of the meson into an outer atomic orbit followed by cascade into the deeper orbits and eventual nuclear interaction. The result of the transitions between atomic orbits is the emission of photons in the X-ray region. The X-ray energies may in many cases be measured to within ten or twenty electronvolts. One may then use the Bohr formula, or the relativistic equivalent, to obtain the meson mass. This yields, for instance, a muon mass of $105\cdot66\{^{+0\cdot02}_{-0\cdot01}$ MeV/c^2.

2.3.3 Decay of charged π- and μ-mesons

We have already seen that the charged π-meson decays into a μ-meson plus at least one additional neutral particle required for energy and momentum conservation. The energy of the μ-meson for π-mesons decaying at rest, or when evaluated in the centre-of-mass system for the moving π-meson, is found to be always the same, 4·1 MeV, indicating that only one additional particle can be present in the decay. Since the masses of the π- and μ-mesons are known, it is possible to calculate the rest mass of the neutral particle; this is found to be consistent with zero. The only known particles of zero rest mass are the photon and the neutrino. Since the direction of the neutral particle is known, lying opposite to the μ-meson in the c.m.s. of the decay, it is possible to search for conversion electrons which may be present if the neutral object is a photon. Such an experiment was done in nuclear emulsions by O'Ceallaigh in 1950, who examined the inferred neutral paths for a number of decays. No electron pairs were observed in a length for which the probability that the gamma ray should not convert was 4×10^{-3}, indicating that the neutral was indeed a neutrino. This conclusion has been verified in later work. Thus we have

$$\pi^{\pm} \rightarrow \mu^{\pm} + \nu.$$

For the π^{+} meson the lifetime may be obtained by a calculation of the proper times for tracks observed in emulsions. More precise determinations have been made using a beam of π-mesons from an accelerator. The beam is stopped in a scintillator in which the pulse from the decay muon may also be observed. The latter method has yielded a value of $(2 \cdot 56 \pm 0 \cdot 05) \times 10^{-8}$ s. Negative π-mesons, when brought to rest, are always absorbed before they can decay. However the lifetime for either charge may be determined by observations of decay in flight, using a beam of π-mesons. For the decay of the μ-meson the energy of the observed decay electron is not unique, but is distributed through a spectrum of values, so that we infer that more than one neutral particle is present. The spectrum indicates that the total mass of the neutrals is again consistent with zero, and no gamma rays are observed to be associated with the decay. The simplest assumption consistent with these observations is for the decay process

$$\mu^{\pm} \rightarrow e^{\pm} + \nu + \bar{\nu}.$$

By methods similar to those described for the pion, the muon lifetime has been found to be $(2 \cdot 203 \pm 0 \cdot 004) \times 10^{-6}$ s. As we shall see in Chapter 4, both of these decay processes are characteristic, in their products and lifetimes, of 'weak' interactions.

Other decay modes are also possible for the charged pion such as

$$\pi \rightarrow \begin{cases} e\nu \\ \mu\nu\gamma \\ e\nu\gamma \\ \pi^{0}e\nu. \end{cases}$$

Only the first two of these have been certainly observed, with frequencies of $(1 \cdot 24 \pm 0 \cdot 03) \times 10^{-4}$ and $(1 \cdot 24 \pm 0 \cdot 25) \times 10^{-4}$ respectively relative to $\pi \rightarrow \mu\nu$.

2.4 The neutral π-mesons

We have already seen that the charge independence of nuclear forces demanded the existence of neutral, as well as charged, mesons. For such a neutral particle the most probable decay consistent with the conservation laws is into two gamma rays.

Such a decay is an electromagnetic process and, as we shall see later, electromagnetic interactions are much stronger than the weak processes; we would thus expect this mode to dominate over possible weak modes, and to have a lifetime considerably shorter than that for the weak decay processes of the charged mesons.

A number of workers had suggested that the π^0 meson might be the source of the soft component of the cosmic radiation, its decay gamma rays giving rise to the showers of photons and electrons. In particular it had been noted that there was a correlation between hard and soft showers. More direct evidence came in 1950 from two experiments, one by Bjorklund et al. studying high-energy photons from a cyclotron target, and the other by Carlson, Hooper and King, using cosmic-ray 'stars' (interactions with secondary prongs in nuclear emulsions). In the experiment of Bjorklund et al., gamma rays arising from the target of the Berkeley cyclotron were allowed to pass through holes in the shielding wall and to enter a magnetic spectrometer. The pairs which they produced in a thin sheet of tantulum were bent in the magnetic field, and detected by proportional counters in coincidence. Gamma-ray spectra were measured for protons incident on the cyclotron target at various energies between 175 and 340 MeV, and for angles of emission from the target of $0°$ and $180°$. A number of possible sources of these gamma rays, which ranged in energy up to 200 MeV for 350 MeV incident protons, were considered. The spectra at $0°$ and $180°$ differed only by a Doppler shift, and became identical when transformed into a coordinate system moving with a velocity of $0.32c$. This corresponds to the expected centre-of-mass velocity for a collision between a 340 MeV proton and a nuclear nucleon, moving in the opposite direction, with an energy of about 25 MeV and a velocity of about $0.2c$. This observation, and also the high energy of the gamma rays, makes it impossible that they should be due to any nuclear excitation. The absolute cross-section for gamma-ray production in the energy range studied is of the order of 10^{-25} mm^2. For bremsstrahlung by protons of this energy we would expect a cross-section of only 10^{-27} mm^2. Also the spectrum shape, and the rather rapid increase in gamma-ray production with increasing incident proton energy, are in complete disagreement with what is to be expected for bremsstrahlung, which cannot therefore explain the origin of the photons.

On the other hand, the hypothesis that the photons arise from the decay into two gamma rays of a neutral meson is found to give a good account of all the features of the observations. A mass of about 150 MeV/c^2 is required for the neutral pion. From this experiment its lifetime can only be said to be less than 10^{-11} s.

The data from the nuclear emulsion work of Carlson, Hooper and King was even more compelling. In this experiment the region around 'stars' formed by cosmic radiation when the emulsions were flown by balloon at a height of twenty-one kilometres was examined for electron–positron pairs. The energy of the tracks of the pairs was determined by measurements of their multiple scattering. The direction of the parent gamma ray could be fixed to within $0.2°$. An analysis of the gamma-ray spectrum deduced from these measurements was then made on the assumption that these gamma rays arose from the decay of π^0 mesons. The fit to

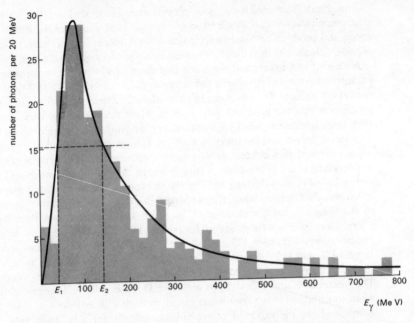

Figure 24 Spectrum of the γ-radiation at 21 km. The full line gives the best fit and the mass of the π^0 meson is deduced from values of E_1 and E_2, such as those shown, according to the equation below (Carlson, Hooper and King, 1950)

the observed spectrum was very good (Figure 24), and the mass of the π^0 could be obtained from the relation

$$\sqrt{(E_1 E_2)} = \tfrac{1}{2} m_0 c^2,$$

where m_0 is the meson mass and E_1 and E_2 are any two values of the gamma-ray energy which correspond to a given number of photons in the spectrum (Figure 24). This yielded a mass for the neutral π-meson of 150 ± 10 MeV/c^2. In confirmation of the interpretation of the photons as arising from π^0 decay, the π^0 spectrum was found to be the same as for charged pions emitted from these stars.

In this experiment it was also possible to set a somewhat lower limit on the life-time of the neutral pion. We have seen that it was possible to fix the direction of a decay gamma ray, which produced an electron–positron pair, within 0·2°. If the π^0 has travelled a short distance from the star before decaying, the line of flight of the gamma ray will not pass through the star (see Figure 25). The distribution of the shortest distance between the line of flight and the star vertex is then a function of the π^0 energy spectrum, which is known, and of the π^0 lifetime. We should note that only the effect of the relativistic time dilatation allows a particle with a lifetime as short as that of the π^0 to travel any measurable distance. In fact this experiment

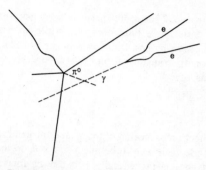

Figure 25 In a cosmic-ray star the line of flight of the γ-ray producing an electron pair may not pass through the vertex due to the finite lifetime (lengthened by time dilatation) of the π^0

could only place an upper limit on the π^0 lifetime, which was however as short as 5×10^{-14} s.

Since this early work the existence of the π^0 meson has been confirmed in a multitude of experiments, perhaps the most striking of which are those where π^0s are produced in reactions in bubble chambers containing liquids of short radiation length. Liquid xenon for instance has a radiation length of only 35 mm while other liquids, less expensive and more commonly used, have radiation lengths in the 0·1–0·3 m range. In such liquids the probability of both decay gamma rays producing pairs is quite large: for if L_p is the mean free path for pair production, then for a beam of photons the number n surviving after a distance x is given by

$$n = n_0 e^{-x/L_P},$$

where n_0 is the initial number. For radiation loss by an electron we have for the average energy E after a distance x

$$E = E_0 e^{-x/L_R},$$

where E_0 is the initial energy and L_R is known as the *radiation length*. Detailed considerations show that $L_p \simeq 1 \cdot 3 L_R$ so that a low radiation length implies a high probability of pair production.

2.5 Best determination of the mass and lifetime of the neutral π-meson

The methods of the original experiment, described above, depend on the measurement of the absolute energy of one of the decay gamma rays. A more precise value may be obtained for the mass in relation to the mass of the π^- meson if

$$m_{\pi^-} - m_{\pi^0} > m_n - m_p.$$

In this case both the following reactions are possible:

$$\pi^- p \to n\gamma, \qquad\qquad\qquad\qquad\qquad\qquad\text{2.2}$$
$$\pi^- p \to n\pi^0, \qquad \pi^0 \to \gamma\gamma. \qquad\qquad\qquad\qquad\text{2.3}$$

We note that the gamma rays from **2.2** for monoenergetic π^- absorption (in practice absorption at rest) will themselves be monoenergetic, while for reaction **2.3** we expect a spectrum of gamma-ray energies. If the mass of the π^0 is $2m_0$ and its velocity is β_0 ($c = 1$), then a Lorentz transformation to the laboratory system gives for the energy of each gamma ray

$$E_\gamma = \frac{m_0 + m_0 \beta_0 \cos \theta}{(1 - \beta_0^2)^{\frac{1}{2}}},$$

where θ is the angle between the direction of motion of the π^0 and the gamma ray, in the π^0 system. The energy dependence is seen to be linear in $\cos \theta$, so that we expect a uniform distribution of gamma-ray energies between E_γ (max) and E_γ (min), where the limiting values occur for gamma rays emitted along the direction, and opposite to the π^0 motion. We can then write

$$\Delta E_\gamma = E_\gamma \,(\text{max}) - E_\gamma \,(\text{min}) = \frac{2m_0 \beta_0}{(1 - \beta_0^2)^{\frac{1}{2}}}$$
$$= P_{\pi^0}.$$

Since the momentum of the neutron is equal to that of the π^0, we can thus obtain the kinetic energy of the neutron. Then writing

$$m_{\pi^-} - m_{\pi^0} = m_n - m_p + T_n + T_{\pi^0},$$

where T_n and T_{π^0} are the neutron and π^0 kinetic energies, and writing T_{π^0} as

$$[m_{\pi^0}^2 + P_{\pi^0}^2]^{\frac{1}{2}} - m_{\pi^0},$$

the values of T_n and P_{π^0} obtained from the gamma-ray spectrum can be used to give $m_{\pi^-} - m_{\pi^0} = 5 \cdot 4 \pm 1 \text{ MeV}/c^2$.

Concerning the π^0 lifetime, more accurate measurements have been made using a sample of decays into one gamma ray plus an electron pair (Dalitz pair). This decay mode occurs in only 1·2 per cent of all cases, but the decay point of the π^0 is defined by the origin of the pair. The distances involved, however, are so small that the precision is still poor.

The most precise value for the π^0 lifetime has been determined indirectly by means of the *Primakoff effect*. This name is given to the process described by the following diagram (Figure 26a) where the π^0 is produced by interaction of an incident gamma ray with a virtual gamma ray of the Coulomb field of a nucleus. It is clear that the interaction described by the function η shown in the diagram is the same as that involved in the π^0 decay (Figure 26b). Thus it is possible that a measurement of the Primakoff-effect cross-section may yield data from which the π^0 lifetime may be obtained. There are certain experimental and theoretical difficulties associated with this procedure. First the cross-section for the Primakoff effect is expected to be quite small (of the order of 1 mb) and very forward peaked. Moreover, in addition to production due to the Primakoff effect, there will also be present both coherent and incoherent π^0 photoproduction due to other nuclear effects. The Primakoff effect is identified by its characteristic angular behaviour,

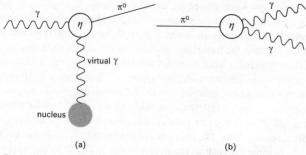

Figure 26 (a) Illustration of the Primakoff effect where a π^0 is produced by the interaction of an incident γ-ray with the electromagnetic field of the nucleus. (b) Illustration of how the π^0 decay arises from the same kind of interaction

and the observed differential cross-section is then fitted as a sum of Primakoff and nuclear processes.

In the experiment of Bellettini *et al.* (1965) the π^0 photoproduction in lead, at 1 GeV gamma-ray energy, was studied around the forward direction. The decay gamma rays were detected by Cerenkov counters in coincidence, each counter subtending a small solid angle of 4×10^{-3} steradians. Scintillation counters in front of

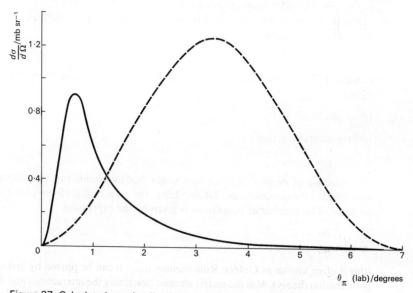

Figure 27 Calculated angular distributions for production of π^0 mesons by the Primakoff process (solid line) and nuclear production (dotted line) (Belletini *et al.*, 1965)

the Cerenkov counters were placed in anticoincidence. Ten different counting channels were mounted on a high precision frame and operated simultaneously. From the expected distributions, shown in Figure 27, for the Primakoff and nuclear production it is clear that the important angular region is entirely below about 7°. Pulse heights from the coincidence counters were displayed on an oscilloscope, and more than one million pulses were photographed. π^0 decays were unambiguously identified from the gamma-ray energies and the requirement for a close coincidence in time. The observed angular distributions could be well fitted *only* by a combination of the Primakoff and nuclear production, and provided a measure of the Primakoff-effect cross-section. The effect of the nuclear production was calculated with certain free parameters such as the nuclear radius. The fit to the data yielded values of these parameters as well as the value of the Primakoff cross-section, and thus of the π^0 lifetime. The Primakoff cross-section is proportional to the reciprocal of the lifetime, and the best fit to all the data yielded a value of $(0 \cdot 73 \pm 0 \cdot 1) \times 10^{-16}$ s. The value obtained for the quantity r_0, in the formula $R = r_0 A^{\frac{1}{3}}$ for the nuclear radius, was $1 \cdot 02 \pm 0 \cdot 07$ fm, in satisfactory agreement with the values obtained for this quantity by other methods.

2.6 The spin of the charged meson

We will discuss the determination of the pion spin both for the reasons of the intrinsic importance of the quantity and also because of the interest of the method of analysis. We shall work always in the c.m.s.

For the π^+ particle the spin is determined by applying the principle of *detailed balance* to the reaction

$$\text{pp} \rightarrow \pi^+ \text{d} \qquad\qquad \textbf{2.4}$$

(which, at the energies under discussion, occurs approximately as frequently as the reaction

$$\text{pp} \rightarrow \text{pn}\pi^+)$$

and the inverse reaction

$$\pi^+ \text{d} \rightarrow \text{pp}. \qquad\qquad \textbf{2.5}$$

The principle of detailed balance then states that the number of transitions per unit time for the two reactions, **2.4** and **2.5**, at the same centre-of-mass energy will be equal. This number of transitions is given by the expression

$$\frac{2\pi}{\hbar} |M|^2 \frac{dn}{dE}$$

(this is often known as Golden Rule number one – it can be proved by first-order perturbation theory). M is the matrix element describing the interactions and dn/dE is the energy density of the final states. Often M can be taken to be independent of the momenta of the final particles; the argument here will not depend on any such special assumptions, but only on the result of the detailed balance theorem which implies also that

$$M_{\pi^+d \to pp} = M_{pp \to \pi^+d}.$$

Normally we are not interested in selecting a particular spin for the final state, and indeed most experiments will not measure this quantity. Thus in evaluating the transition probability T we will sum over all the final spin states and write

$$T = \frac{2\pi}{\hbar} \sum_f |M_f|^2 \frac{dn}{dE},$$

where the summation is over all final states f.

This gives the total transition rate from *all* initial spin states, but each reaction proceeds from a particular spin configuration so that we must use an *average* of the transition probabilities from all the initial states, which will be all equally probable. For each of the initial particles of spin s there are $(2s+1)$ spin states so that for the two initial particles having spins s_1 and s_2 we have altogether $(2s_1+1)(2s_2+1)$ initial states. Then the average over all states for the transition probability is

$$\frac{2\pi}{\hbar} [(2s_1+1)(2s_2+1)]^{-1} \sum_f |M_f|^2 \frac{dn}{dE}.$$

The cross-section is related to the transition probability in the following way:

Transition probability = cross-section $(\sigma) \times$ number of particles/unit volume $\times$
$\times$ relative velocity of interacting particles (v),

so that

$$\sigma_\theta = [(2s_1+1)(2s_2+1)]^{-1} \sum_f |M_f|^2 \frac{dn}{dE} \text{(unit vol./no. of particles)}^{-1} v^{-1}.$$

For a single particle in a volume V we have

$$\sigma_\theta = [(2s_1+1)(2s_2+1)]^{-1} \sum_f |M_f|^2 \frac{dn}{dE} V . v_i^{-1},$$

where v_i is the relative velocity in the initial state.

Now we must consider the dn/dE factor. Here we refer to a result familiar from statistical mechanics, that the number of states δn for a particle with momentum in the interval between p and $p+\delta p$, in a box of volume V, is

$$\delta n = 4\pi p^2 \delta p \frac{V}{(2\pi\hbar)^3},$$

so that
$$\frac{dn}{dp} = \frac{4\pi p^2 V}{(2\pi\hbar)^3}.$$

But if ε_1 and ε_2 are the total particle energies then

$$\frac{dE}{dp} = \frac{d\varepsilon_1}{dp} + \frac{d\varepsilon_2}{dp} = v_f \left(p^2 = \varepsilon^2 - m^2, \text{ so } \frac{d\varepsilon}{dp} = \frac{p}{\varepsilon} = v \right),$$

where v_f is the relative velocity of the particles in the final state. Thus

$$\frac{dn}{dE} = \frac{4\pi p^2 V}{(2\pi\hbar)^3 v_f}.$$

Substituting, we obtain an expression for

$$\sigma_\theta = [(2s_1+1)(2s_2+1)]^{-1} \sum_f |M_f|^2 \frac{4\pi p^2 V^2}{(2\pi\hbar)^3 v_i v_f}.$$

Now at the same centre-of-mass energy, using the detailed balancing result that the matrix elements in both directions are equal, we write

$$M_{\pi^+ d \to pp} = M_{pp \to \pi^+ d}.$$

Thus we have

$$\sigma_{\pi^+ d \to pp}(2s_{\pi^+}+1)(2s_d+1)v_{\pi^+ d} v_{pp}(p_p)^{-2} = \sigma_{pp \to \pi^+ d}(2s_p+1)^2 v_{\pi^+ d} v_{pp}(p_{\pi^+})^{-2},$$

so that

$$\sigma_{\pi^+ d \to pp}(2s_\pi+1)3 = \sigma_{pp \to \pi^+ d} 4\left(\frac{p_p}{p_{\pi^+}}\right)^2$$

and

$$\sigma_{pp \to \pi^+ d} = \sigma_{\pi^+ d \to pp} \frac{3}{4}(2s_\pi+1)\left(\frac{p_\pi}{p_p}\right)^2,$$

at any angle θ. Integrating over all angles, a factor of 2 must be introduced to take account of the indistinguishability of the two protons in the final state in the $\pi^+ d$ absorption reaction giving

$$\sigma_{pp \to \pi^+ d} = 2\sigma_{\pi^+ d \to pp} \frac{3}{4}(2s_\pi+1)\left(\frac{p_\pi}{p_p}\right)^2.$$

This last equation can now be used to determine the spin of the π^+.

We note first that even rather approximate values of the two cross-sections will yield the value of the spin, since we already know that this must be an integer. If we use the original data (Cartwright *et al.*, 1953) for the reaction $pp \to \pi^+ d$ from experiments carried out at Berkeley at an incident proton energy of 340 MeV, we have a cross-section of 0.18 ± 0.06 mb. This energy for the incident proton corresponds to a meson energy of 22.3 MeV in the c.m.s. For the reaction $\pi^+ d \to pp$ the cross-section for π^+ mesons incident on deuterium, and having an energy of 29 MeV, is 3.1 ± 0.3 mb (Durbin *et al.*, 1951). This incident energy corresponds to a meson energy of 25 MeV in the c.m.s., sufficiently close to that in the proton–proton work mentioned above. To obtain the value of the ratio p_π/p_p we may use the non-relativistic relationships, since the energies are quite low. For a kinetic energy of 23 MeV for the pion, and a corresponding energy of 85 MeV for each proton in the c.m.s., we obtain a ratio $p_\pi/p_p = 0.20$. Substituting in the equation we then have

$$(0.18 \pm 0.06) = (3.1 \pm 0.3) \times 1.5 \times 0.04(2s_\pi+1).$$

Thus $(2s_\pi + 1) = 1$ and the pion spin is 0. Data on these reactions at other energies have confirmed this conclusion. A further indication of zero spin comes from the observation that the decay of charged mesons is found to be closely isotropic in the c.m.s. We will assume that the spin of the negative pion is the same as for the positive pion. This assumption is found to lead to consistent interpretations of a large variety of processes involving such particles. Discussion of the spin of the neutral pion is delayed until we have dealt with the subject of parity. We should of course expect that its spin also would be the same as for the charged pions.

We may note that our analysis depended on having an incident pion beam for the π^+p absorption in which all spin states were equally probable. If in fact the pion spin was not zero, *and* if the pions in the beam were polarized, then this assumption would be invalid. It is, however, highly unlikely that the beam would be substantially polarized, and in such a way as to simulate the result for a beam of spin-zero particles.

2.7 Résumé concerning parity and the parity of the charged π-meson

We shall deal with the significance of conservation laws in general in the following chapters. In order to complete the discussion of the pion properties, however, it is convenient to discuss the pion parity here. Such a discussion should also lend some degree of reality to the subsequent more general treatment.

The parity operator reverses each of the coordinate axes. That is it produces the transformation

$$x \to -x,$$

$$y \to -y,$$

$$z \to -z.$$

If a wave function subjected to this transformation is unchanged, then it is said to have positive parity, i.e.

$$P\psi(x, y, z) = \psi(-x, -y, -z) = +1\psi(x, y, z),$$

where P is the parity operator producing the above transformation. Alternatively we may have

$$P\psi(x, y, z) = \psi(-x, -y, -z) = -1\psi(x, y, z).$$

In the first case we say that the parity of the wave function is even, or sometimes we say it is $+1$ meaning that the eigenvalue of the wave function under P is $+1$, while in the second case we say that the parity is odd or -1. All physically meaningful wave functions must fall into one or other category since $|\psi|^2$ must be invariant under the parity transformation. This property is clearly related to the symmetry of the wave function as is illustrated in Figure 28. Invariance of a system under the parity transformation implies that the eigenvalues of the wave function under this transformation are conserved quantities.

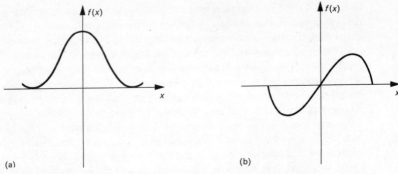

Figure 28 The function in (a) is *even* under inversion of x while that in (b) is *odd*

We may look on the parity transformation as consisting of two steps: (a) rotation by $180°$ around the x-axis giving

$y \rightarrow -y$,

$z \rightarrow -z$;

(b) reversal of the x-axis

$x \rightarrow -x$,

that is a $180°$ rotation followed by a 'mirror image' transformation. As we shall see, conservation of angular momentum implies invariance under rotations so that the *additional* invariance implied by the parity transformation is invariance under mirror imaging.

In practice of course we are always concerned with systems consisting of at least two particles. In such a system the parity of the whole wave function can be considered as the product of the 'intrinsic' parities of the individual particles and the parity of the orbital angular-momentum part of the wave function. In this we simply state that we can represent, for instance, the spatial wave function of a pion and a proton in the form

$\psi(p)\, \psi(\pi)\, \psi$(orbital angular momentum L of p and π),

and then

$P(\pi - p) = P(p)\, P(\pi)\, P$(orbital angular-momentum wave function).

Considering the orbital angular-momentum part we need only refer to the spherical harmonics required to describe the appropriate wave functions for s, p, d, ... states to see that the parity is given in terms of the orbital angular-momentum quantum number L by $P(L) = (-1)^L$. For instance,

$$\ldots \text{for the} \begin{Bmatrix} \text{s-state} \\ \text{p-state} \\ \text{d-state} \end{Bmatrix} \psi\,(\text{orbital}) \begin{cases} \propto \text{constant} & L = 0 \quad \text{parity is } +1 \\ \propto \cos\theta & L = 1 \quad \text{parity is } -1 \\ \propto (3\cos^2\theta - 1) & L = 2 \quad \text{parity is } +1. \end{cases}$$

The intrinsic parities of particles, as we shall use this quantity, have meaning only relative to each other. We *choose* the parity of the proton to be positive. If the parity of the neutron is also positive then the parity of the deuteron, according to our rule described above, will be positive. Indeed a knowledge of the deuteron wave function may be used to establish the neutron parity as positive.

We may now return to the question of the parity for the charged mesons. Consider the processes

$$\pi^- d \rightarrow nn, \tag{2.6}$$

$$\pi^- d \rightarrow nn\gamma, \tag{2.7}$$

$$\pi^- d \rightarrow nn\pi^0. \tag{2.8}$$

We must consider first the angular momentum state of the $\pi^- - d$ system when the meson is absorbed at rest. In fact it has been shown that the time for a π^- to reach the K-orbit in a $\pi^- d$ atom is only $\sim 10^{-10}$ s. Also, direct nuclear capture of the π^-, even from the 2p-level, is negligible compared with the 2p to 1s transition. Thus all π^-s will be captured from the s-state in the $\pi^- d$ atom before they can decay. This means that the parity of the initial state is simply the parity of the pion, while the total angular momentum of the initial state is 1.

For the reaction **2.6** above, the two-neutron state must be a p-state since the Pauli principle forbids a $^3s_1$ state. Thus the final state has odd parity and, assuming that the parity is conserved in the reaction (we shall see later that there is very good evidence that parity is conserved in all 'strong' reactions), the process **2.6** can take place only if the pion parity is odd.

The reaction **2.6** has been clearly demonstrated to take place by direct observation of the neutrons, which have a unique energy for this process. The ratio of **2.6** to **2.7** equals $2 \cdot 35 \pm 0 \cdot 35$. The reaction **2.8** is not observed, as is established by the absence of decay gamma rays from the neutral pion which would have an energy of about 70 MeV. The fact that this reaction does not take place is not surprising since only 2·3 MeV is available as particle kinetic energy in the final state and, since an $L = 1$ combination is required to give a total angular momentum of 1, the process is very unlikely. The reaction also turns out to be forbidden by parity conservation as we shall see later.

The pion thus has spin–parity 0^- and is called pseudo-scalar particle.

2.8   Spin and parity of the neutral π-meson

For two-photon decay the photons must have equal and opposite momenta in the centre-of-mass system. No generality is lost if we consider both photons as circularly polarized and it will be convenient to do this in the first instance. Thus the final two-photon system can take the four forms shown in Figure 29. We use the letters *l* and *r* to signify left and right circular polarization for the photon travelling in the $-z$ direction and *L* and *R* as the equivalent symbols for the $+z$ direction, where we have defined the z-axis as the direction of propagation. With this definition the photons will always have $J_z = \pm 1$ regardless of the total angular

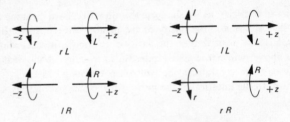

Figure 29 The four possible polarization states for a two-photon system

momentum J of the decaying particle. Thus for the two photons the total z-component of the angular momentum is

$$J_z = \pm 2 \quad \text{for } rL \text{ and } lR: \text{ parallel spins}$$

and $\quad J_z = 0 \quad$ for lL and rR: opposed spins.

Since angular momentum is conserved, rL and lR may occur only for $J \geqslant 2$.

We now consider the following functions: rR, lL, rL, lR. We will examine their behaviour under transformations which should leave the system invariant. First we take a rotation of 180° about the x-axis so that the $+z$ and $-z$ directions are interchanged. This represents simply an interchange of the two gamma rays, which must leave the system as a whole unchanged. Then all angles θ measured from $+z$ become $(\pi - \theta)$, $l \leftrightarrow L$ and $r \leftrightarrow R$. This transformation conserves J and J_z. We may write the angular dependence of the initial and final states in terms of spherical harmonics $Y_J^{J_z}(\theta)$. Spherical harmonics moreover have the general property

$$Y_J^{J_z}(\pi - \theta) = (-1)^J \, Y_J^{J_z}(\theta).$$

It is then clear that J cannot be equal to 1 since in this case the system would not be invariant under the transformation in question. Thus two-photon decay is forbidden for vector or pseudo-vector particles.

Now let us consider the effect of the parity transformation. We have already seen that this corresponds to a rotation of 180° about the z-axis plus inversion of the z-axis. The result is therefore $R \leftrightarrow l$, $r \leftrightarrow L$. This means that the two-photon wave functions transform as follows:

$$Rr \leftrightarrow lL, \qquad Rl \leftrightarrow lR, \qquad Lr \leftrightarrow rL, \qquad Ll \leftrightarrow rR.$$

Thus for the parity transformation the two-quantum wave functions Rr and Ll are *not* eigenfunctions. However the combinations $Rr + Ll$ and $Rr - Ll$ are eigenfunctions. We will consider these functions for which zero is the lowest possible spin value, and also the corresponding functions $Rl + Lr$ and $Rl - Lr$ where two is the lowest possible value for the spin. Applying the parity transformation to these functions we see that

$$0_+ \leftrightarrow 0_+$$
$$0_- \leftrightarrow -0_-$$
$$2_+ \leftrightarrow 2_+$$
$$2_- \leftrightarrow 2_-,$$

where we have written 0_+ and 0_- for the symmetrical and antisymmetrical wave functions corresponding to lowest spin 0, and 2_+ and 2_- for the corresponding wave functions for lowest spin 2.

Therefore for systems of even parity only 0_+, 2_+, and 2_- photon wave functions can occur while for odd parity only the 0_- function can occur.

If we reject the higher spins 2, 4, 6, ..., which lead to trouble in the field theory, and take the lowest value consistent with the data, we thus get the same spin for the neutral as for the charged pions, as we might expect. The parity is then seen to be allowed to be either $+$ or $-$ according to the above argument. A distinction between these two possibilities can be made by measurement of the polarization of the decay gamma rays.

In order to see what is to be expected in the two possible cases, it is convenient to rewrite the wave functions in terms of plane-polarized photons. We may represent the circularly-polarized photon by a sum of two wave functions with appropriate relative phase, one of which represents a photon plane-polarized in the x-direction and the other a photon plane-polarized in the y-direction. Thus we write

$$L = P_x + iP_y = R^* \quad \text{and} \quad l = p_x - ip_y = r^*,$$

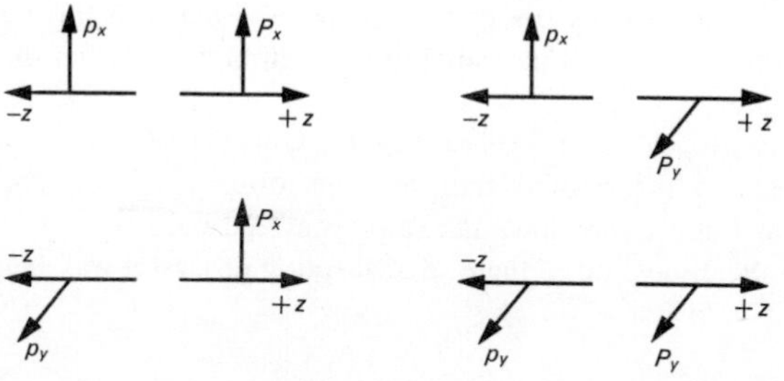

Figure 30 Plane-polarization representations for a two-photon state

where P and p are real. The various cases are symbolized in Figure 30. Then in terms of the previous photon wave functions we can write

$$0_+ = lL + rR$$
$$= (p_x - ip_y)(P_x + iP_y) + \text{complex conjugate}$$
$$= 2\,\text{Re}(lL)$$
$$= 2(p_x P_x + p_y P_y).$$

Similarly

$$2_+ = 2(p_x P_x - p_y P_y),$$
$$0_- = -2i(p_x P_y - p_y P_x),$$
$$2_- = -2i(p_x P_y + p_y P_x).$$

If the pion spin is zero (threshold behaviour of π^0 production is certainly consistent with this hypothesis, although it cannot prove absolutely that it is true), then this last result affords a means of determining whether the π^0 is a scalar or pseudo-scalar particle. For *even* parity the planes of polarization of the decay photons are parallel while for *odd* parity they are perpendicular.

The polarization of high-energy photons such as those from the π^0 decay is difficult to measure. However the π^0 decay into two electron pairs affords the possibility of measuring the polarization as a correlation between the planes of pairs. Such a correlation has been studied by Plano et al. (1959). The double-pair decay is very rare, occurring in only 1 in 30 000 decays. In the experiment in question, π^0 mesons were produced by the capture reaction

$$\pi^- p \to \pi^0 n.$$

About one-and-a-half million stopped π^- yielded about two hundred double-pair events. The theoretical correlation then has the form

$$I_0^+ (\theta) = 1 + A \cos 2\theta \qquad \text{scalar } \pi^0,$$
$$I_0^- (\theta) = 1 - A \cos 2\theta \qquad \text{pseudo-scalar } \pi^0,$$

where θ is the angle between the planes of the two pairs and the coefficient A is a function of the angles and energy division in the pairs. Even with this relatively small number of events the results indicated quite unambiguously that the parity is odd.

A further evidence for odd parity comes from the considerable data in support of charge independence which requires that the pions form an isotopic-spin triplet, the members of which must then have the same spin and parity.

We recall that in the discussion of the $\pi^- d$ absorption process it was mentioned that the process

$$\pi^- d \to n n \pi^0$$

turned out to be forbidden by parity conservation. It was shown there that the initial state has the parity of the π^-, i.e. odd, while the final state involved orbital angular momentum $L = 1$. Thus the π^0 parity P_{π^0} would be given by

$$P_{\pi^0} \cdot + \cdot - = -, \quad \text{i.e.} \quad P_{\pi^0} = +,$$

if the reaction was observed to take place. This involves the assumption that the π^0 is in an s-state with respect to the n–n pair.

2.9 Isotopic spin for the π-meson

As with parity we shall discuss this question from a more general point of view in Chapter 3. However it will be useful to have available the example of the π-nucleon system, so that we will deal briefly with the idea and application of isotopic spin, i-spin, at this point. We will not develop here the complete i-spin formalism.

We first recall that in all properties except those associated with electric charge the neutron and proton are practically identical. It is therefore reasonable to denote these particles as two orientations of a vector in a space which we call charge space or isotopic-spin space. The charge of the particle depends only on the orientation of the vector. The reason for the i-spin nomenclature is thus clear in that there is a precise analogy between this idea and the concept of ordinary spin. A particle having ordinary spin $\frac{1}{2}$ has two possible orientations, $+\frac{1}{2}$ and $-\frac{1}{2}$, for its angular-momentum vector. Similarly the nucleon has two possible states of orientation of the isotopic-spin vector, which are manifest as the proton and the neutron.

In angular-momentum theory it is shown that the number of states for a system with angular momentum J is $2J+1$. The development of the theory for i-spin is identical, so that for the nucleon the number of states, which we know to be two, shows that the i-spin must be $\frac{1}{2}$. It is natural then to assign the neutron as the state with $I_3 = -\frac{1}{2}$ and the proton to $I_3 = +\frac{1}{2}$, where I_3 is the 'third' component of the isotopic spin. In this case the charge Q is seen to be related to I_3 by

$$Q = I_3 + \tfrac{1}{2}.$$

At this stage of course we have learned nothing new but have merely introduced an alternative notation for nucleons. This notation facilitates some interesting conclusions in nuclear physics, discussed in other volumes, but here we proceed to deal with the i-spin of π-mesons before attempting to illustrate the usefulness of the concept in situations with two or more particles.

We know that there exist three π-mesons which are almost identical except for their charges. Thus we write $2I_\pi + 1 = 3$, so that $I_\pi = 1$ and we assign

$$(I_{\pi^-})_3 = -1, \qquad (I_{\pi^0})_3 = 0, \qquad (I_{\pi^+})_3 = +1.$$

Note that the charge–i-spin relationship is now apparently different from that for nucleons in that $Q = I_3$. This is one of the first symptoms that mesons are fundamentally different from nucleons. The situation is rationalized in an *ad hoc* manner by assigning a 'baryon number' (i.e. heavy particle number) B of $+1$ to the nucleons and 0 to the pions. Then we can write for nucleons *or* mesons

$$Q = I_3 + \tfrac{1}{2}B.$$

Now we make the important hypotheses:

(a) Strong interactions (those involving nucleons and mesons only) are independent of I_3, and thus of Q, and depend only on the total i-spin, I.

(b) The total i-spin is conserved in strong interactions; i.e. I is a good quantum number in such processes.

We shall justify these hypotheses by comparison of the results obtained by their use with experiments. We note of course that I_3 must also be conserved since Q and B are both conserved quantities

2.10 Pion–proton scattering

The study of pion–proton scattering provides a first and most striking illustration of the usefulness of the i-spin concept.

We first consider all the possible i-spin states of the pion–nucleon system. The procedure is the same as for ordinary spin so that the total i-spin can either be $I = \frac{3}{2}$ or $I = \frac{1}{2}$. Thus the range of states for the system is:

$$(I, I_3) = (\tfrac{3}{2}, \tfrac{3}{2}); \quad (\tfrac{3}{2}, \tfrac{1}{2}); \quad (\tfrac{3}{2}, -\tfrac{1}{2}); \quad (\tfrac{3}{2}, -\tfrac{3}{2}); \quad (\tfrac{1}{2}, \tfrac{1}{2}); \quad (\tfrac{1}{2}, -\tfrac{1}{2}). \tag{2.9}$$

The first hypothesis then asserts that the first four states of **2.9** behave identically to each other as far as strong interactions are concerned, and that the latter two states also behave identically for strong interactions.

Now we must write these states in terms of the i-spin wave functions of the pions and the nucleons. The notation will be that the i-spin wave functions will be written as

$$\left(I_{\text{pion}}, [I_{\text{pion}}]_3\right)\left(I_{\text{nucleon}}, [I_{\text{nucleon}}]_3\right),$$

while the left-hand side of the equation will give the wave function for the total state in the corresponding notation. For the $(\tfrac{3}{2}, \tfrac{3}{2})$ state we have only one possible pion–nucleon combination: $(1, 1)(\tfrac{1}{2}, \tfrac{1}{2})$. For the state $(\tfrac{3}{2}, -\tfrac{3}{2})$ the only possible combination is $(1, -1)(\tfrac{1}{2}, -\tfrac{1}{2})$. For the other four states two possible pion–nucleon combinations can yield the required state which is thus expressed as the sum of the combinations, weighted with the appropriate probabilities which are given by the relevant Clebsch–Gordan coefficients (see Appendix B). Thus we have

$$(\tfrac{3}{2}, \tfrac{1}{2}) = \sqrt{\tfrac{1}{3}}(1, 1)(\tfrac{1}{2}, -\tfrac{1}{2}) + \sqrt{\tfrac{2}{3}}(1, 0)(\tfrac{1}{2}, \tfrac{1}{2}),$$
$$(\tfrac{3}{2}, -\tfrac{1}{2}) = \sqrt{\tfrac{2}{3}}(1, 0)(\tfrac{1}{2}, -\tfrac{1}{2}) + \sqrt{\tfrac{1}{3}}(1, -1)(\tfrac{1}{2}, \tfrac{1}{2}),$$
$$(\tfrac{1}{2}, \tfrac{1}{2}) = \sqrt{\tfrac{2}{3}}(1, 1)(\tfrac{1}{2}, -\tfrac{1}{2}) - \sqrt{\tfrac{1}{3}}(1, 0)(\tfrac{1}{2}, \tfrac{1}{2}),$$
$$(\tfrac{1}{2}, -\tfrac{1}{2}) = \sqrt{\tfrac{1}{3}}(1, 0)(\tfrac{1}{2}, -\tfrac{1}{2}) - \sqrt{\tfrac{2}{3}}(1, -1)(\tfrac{1}{2}, \tfrac{1}{2}).$$

We can now replace the brackets on the right by the names of the particles themselves, in order to exhibit more clearly the nature of the wave functions, although it should still be remembered that in this context the names represent the wave functions. Rewriting in this way we have

$$(\tfrac{3}{2}, \tfrac{3}{2}) = \pi^+ p,$$
$$(\tfrac{3}{2}, \tfrac{1}{2}) = \sqrt{\tfrac{1}{3}}\pi^+ n + \sqrt{\tfrac{2}{3}}\pi^0 p,$$
$$(\tfrac{3}{2}, -\tfrac{1}{2}) = \sqrt{\tfrac{2}{3}}\pi^0 n + \sqrt{\tfrac{1}{3}}\pi^- p, \tag{2.10}$$
$$(\tfrac{3}{2}, -\tfrac{3}{2}) = \pi^- n,$$

and
$$(\tfrac{1}{2}, +\tfrac{1}{2}) = \sqrt{\tfrac{2}{3}}\pi^+ n - \sqrt{\tfrac{1}{3}}\pi^0 p,$$
$$(\tfrac{1}{2}, -\tfrac{1}{2}) = \sqrt{\tfrac{1}{3}}\pi^0 n - \sqrt{\tfrac{2}{3}}\pi^0 p. \tag{2.11}$$

A little algebra then gives the i-spin wave functions for different pion–nucleon combinations

$\pi^+ p = (\frac{3}{2}, \frac{3}{2})$,

$\pi^- p = \sqrt{\frac{1}{3}}(\frac{3}{2}, -\frac{1}{2}) - \sqrt{\frac{2}{3}}(\frac{1}{2}, -\frac{1}{2})$,

$\pi^0 p = \sqrt{\frac{2}{3}}(\frac{3}{2}, \frac{1}{2}) - \sqrt{\frac{1}{3}}(\frac{1}{2}, \frac{1}{2})$,

$\pi^+ n = \sqrt{\frac{1}{3}}(\frac{3}{2}, \frac{1}{2}) + \sqrt{\frac{2}{3}}(\frac{1}{2}, \frac{1}{2})$,

$\pi^0 n = \sqrt{\frac{2}{3}}(\frac{3}{2}, \frac{1}{2}) + \sqrt{\frac{1}{3}}(\frac{1}{2}, -\frac{1}{2})$,

$\pi^- n = (\frac{3}{2}, -\frac{3}{2})$.

We note that the $\pi^+ p$ and $\pi^- n$ states are states of pure i-spin while the other pion–nucleon states consist of mixtures of i-spin-$\frac{3}{2}$ and i-spin-$\frac{1}{2}$ states.

On the basis of our earlier assumption we can now describe all the meson-nucleon scattering processes in terms of only two scattering amplitudes $A(\frac{3}{2})$ and $A(\frac{1}{2})$, compared with eight amplitudes in the case of no relationship between the different processes. We may now write the amplitudes for the processes

$\pi^+ p \to \pi^+ p$,

$\pi^- p \to \pi^- p$,

$\pi^- p \to \pi^0 n$.

We use the result that the cross-section for any process is proportional to the square of the matrix element between initial and final states:

$\sigma \propto |(\psi_f|H|\psi_i)|^2$,

where H is the appropriate matrix element.

Then using the i-spin wave functions as above we obtain

$\sigma(\pi^+ p \to \pi^+ p) \propto |(\frac{3}{2}, \frac{3}{2})H(\frac{3}{2})(\frac{3}{2}, \frac{3}{2})|^2$,

for π^+-proton elastic scattering,

$\sigma(\pi^- p \to \pi^- p) \propto |(\sqrt{\frac{1}{3}}(\frac{3}{2}, -\frac{1}{2})H(\frac{3}{2})\sqrt{\frac{1}{3}}(\frac{3}{2}, -\frac{1}{2})) + (\sqrt{\frac{2}{3}}(\frac{1}{2}, -\frac{1}{2})H(\frac{1}{2})\sqrt{\frac{2}{3}}(\frac{1}{2}, -\frac{1}{2}))|^2$,

where we have used the conservation of total i-spin, for π^--proton elastic scattering, and

$\sigma(\pi^- p \to \pi^0 n) \propto |(\sqrt{\frac{2}{3}}(\frac{3}{2}, -\frac{1}{2})H(\frac{3}{2})\sqrt{\frac{1}{3}}(\frac{3}{2}, -\frac{1}{2})) - (\sqrt{\frac{1}{3}}(\frac{1}{2}, -\frac{1}{2})H(\frac{1}{2})\sqrt{\frac{2}{3}}(\frac{1}{2}, -\frac{1}{2}))|^2$,

for charge-exchange scattering. If we then write

$A(\frac{3}{2}) = (\frac{3}{2}, \ldots)H(\frac{3}{2})(\frac{3}{2}, \ldots)$,

$A(\frac{1}{2}) = (\frac{1}{2}, \ldots)H(\frac{1}{2})(\frac{1}{2}, \ldots)$,

we obtain $\quad \sigma(\pi^+ p \to \pi^+ p) \propto |A(\frac{3}{2})|^2$,

$\sigma(\pi^- p \to \pi^- p) \propto |(\frac{1}{3})A(\frac{3}{2}) + (\frac{2}{3})A(\frac{1}{2})|^2$,

$\sigma(\pi^- p \to \pi^0 n) \propto |(\sqrt{\frac{2}{9}})A(\frac{3}{2}) - (\sqrt{\frac{2}{9}})A(\frac{1}{2})|^2$.

This result enables us to compare cross-sections for the three processes at the same angles and energies. Thus

$$\sigma(\pi^+p \to \pi^+p) : \sigma(\pi^-p \to \pi^-p) : \sigma(\pi^-p \to \pi^0n)$$
$$= \left|A(\tfrac{3}{2})\right|^2 : \left|(\tfrac{1}{3})A(\tfrac{3}{2}) + (\tfrac{2}{3})A(\tfrac{1}{2})\right|^2 : \left|(\sqrt{\tfrac{2}{9}})A(\tfrac{3}{2}) - (\sqrt{\tfrac{2}{9}})A(\tfrac{1}{2})\right|^2.$$

Further progress in predicting the ratios of the cross-sections depends on a knowledge of $A(\tfrac{3}{2})$ and $A(\tfrac{1}{2})$. We examine some special cases:

(a) $A(\tfrac{3}{2}) \gg A(\tfrac{1}{2})$ then $\sigma(\pi^+p \to \pi^+p) : \sigma(\pi^-p \to \pi^-p) : \quad \sigma(\pi^-p \to \pi^0n)$

$$= \qquad 1 \qquad : \qquad \tfrac{1}{9} \qquad : \qquad \tfrac{2}{9}$$

(b) $A(\tfrac{1}{2}) \gg A(\tfrac{3}{2})$ then $\sigma(\pi^+p \to \pi^+p) : \sigma(\pi^-p \to \pi^-p) : \sigma(\pi{-}p \to \pi^0n)$

$$= \qquad 0 \qquad : \qquad \tfrac{4}{9} \qquad : \qquad \tfrac{2}{9}$$

(c) $A(\tfrac{3}{2}) = A(\tfrac{1}{2})$ then $\sigma(\pi^+p \to \pi^+p) : \sigma(\pi^-p \to \pi^-p) : \sigma(\pi^-p \to \pi^0n)$

$$= \qquad 1 \qquad : \qquad 1 \qquad : \qquad 0.$$

Now let us turn to the measured values. The experiments involve the production of a beam of positive or negative mesons, normally by means of a cyclotron. The π-meson beam is focused and analysed in momentum, by passing through a bending magnet followed by a suitable collimator, before being allowed to pass into a target of liquid hydrogen. In the elastic scattering processes the charged meson and the proton are then detected by counter telescopes in coincidence, in which may be measured both the energy and the rate of loss of energy of the particles, so

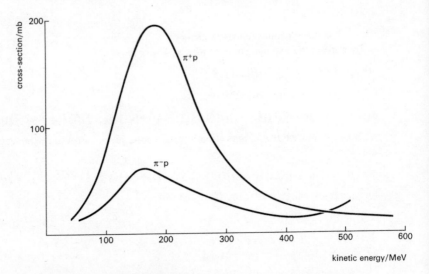

Figure 31 Cross-sections for the scattering of π^+ and π^- mesons on protons showing the Δ (1238) resonance. The π^-p cross-section is the sum of the elastic and charge-exchange data

that they may be identified. In such a two-particle process, with a monoenergetic incident beam, the conservation of energy and momentum impose tight conditions on the energies and angles of the outgoing particles. The elastic scattering events may thus be easily and unambiguously identified, especially in the region where the energy is not too high. In the charge-exchange scattering the π^0 is normally detected by observations on the decay gamma rays. The results for the total cross-section for π^+p and π^-p scattering, as a function of the kinetic energy of the incident pion, are shown in Figure 31. The most striking feature of the data is the strong peak in the cross-section for a pion kinetic energy of about 180 MeV. In the region of this peak, or 'resonance', the cross-section ratios are found to be 195, 23 and 45 mb, in close agreement with the $9:1:2$ ratio, indicating that this resonance arises from an interaction which takes place very predominantly in the $I = \frac{3}{2}$ state. The success of this analysis is a compelling proof of the correctness of the earlier hypotheses. Note that this result is certainly not apparent on the basis of our ordinary ideas concerning charge conservation and charge independence.

We shall return to a fuller discussion of the properties of this $I = \frac{3}{2}$ resonance in Chapter 9.

Chapter 3
Conservation laws

3.1 General features of conservation laws

In general terms we will say that a *theory* gives us the equation of motion for a system. Examples are the Schrödinger equation, the Dirac equation, Maxwell's equations, and the Lagrange equations. To solve a given problem it is necessary to integrate such differential equations of motion, and occasionally this can be done analytically. The equations are generally of second order in time, so that two integrations are necessary for their solution. In many cases the first integral gives a conservation law.

The most familiar of such equations are probably the Lagrange equations. For a system with Lagrangian $L = T - V$ the generalized momenta are given by

$$p_i = \frac{\partial L}{\partial \dot{q}_i}$$

and the Lagrange equations for the system are

$$\frac{d}{dt}\frac{\partial L}{\partial \dot{q}_i} - \frac{\partial L}{\partial q_i} = 0,$$

where q_i are the generalized coordinates and T and V the kinetic and potential energy respectively. Now suppose that a particular coordinate is missing from the Lagrangian of a system. In such a case the Lagrangian is independent of the quantity in question, or symmetrical with respect to transformations of this coordinate. For instance the Lagrangian may be independent of θ, the angle of rotation about the z-axis, for fields symmetrical about this axis. Then

$$\frac{\partial L}{\partial q_i} = 0,$$

and we have $\quad \dfrac{d}{dt}\dfrac{\partial L}{\partial \dot{q}_i} = 0$

or since $\quad \dfrac{\partial L}{\partial \dot{q}_i} = p_i, \qquad \dfrac{dp_i}{dt} = 0.$

so that p_i is a constant. This means that the generalized momentum conjugate to a coordinate, with respect to which the system is symmetrical, is conserved. In the example already mentioned, if we write $q_i = \theta$ and if we take V to be independent of velocity, then for a particle of mass m

$$T = \tfrac{1}{2}m\theta^2 r^2$$

so that $\quad p_i = m\theta r^2$

$$\quad = \text{angular momentum.}$$

Thus the statement that the angular momentum is conserved is equivalent to the statement that the system is invariant with respect to rotations about the relevant axis, or is symmetrical with respect to such an axis. Another way of describing such a symmetry is to say that it implies that there is some aspect of the physical system which is *irrelevant* to its behaviour.

Similarly for the conservation of linear momentum we may write the Lagrangian for motion of a particle of mass m along the x-axis in a potential V which is independent of x as (non-relativistic case)

$$L = \tfrac{1}{2}m\dot{x}^2 - V(y, z)$$

and the momentum $p_x = m\dot{x}$ so that the Lagrange equation gives p_x as constant. Thus we see that if the system is invariant under translation in the x-direction then the component of the momentum in the x-direction is conserved, while the converse is also true.

3.2   Conservation laws in quantum theory

It is useful to note some formal properties of the operators and wave functions of quantum mechanics under transformation.

If $\psi(\mathbf{r}_i)$ is a wave function in system 1, then in system 2 we can write a wave function $\psi'(\mathbf{r}_i)$ which is related to $\psi(\mathbf{r}_i)$ by an operator U such that

$$\psi'(\mathbf{r}_i) = U\psi(\mathbf{r}_i).$$

It can readily be shown that if ψ and ψ' are independently normalized to unity then U is a unitary operator, i.e.

$$U^+U = 1,$$

where U^+ is the Hermitian conjugate of U and

$$U^{-1} = U^+.$$

Now associated with every observable we have an operator Q; for instance with linear momentum along the x-axis we associate $-i\hbar\,\partial/\partial x$. Then the expectation value $\langle Q \rangle$ for the observable is

$$\langle Q \rangle = [\psi(\mathbf{r}_i)|Q|\psi(\mathbf{r}_i)] = \int \psi^* Q\psi \, d\tau.$$

In the transformed coordinate system we can get the expectation value $\langle Q \rangle'$ from the expression

$$\langle Q \rangle' = [\psi'(\mathbf{r}_i)|Q|\psi'(\mathbf{r}_i)]$$

and using the relationship between $\psi(\mathbf{r}_i)$ and $\psi'(\mathbf{r}_i)$

$$\langle Q \rangle' = [U\psi(\mathbf{r}_i)|Q|U\psi(\mathbf{r}_i)]$$
$$= [\psi(\mathbf{r}_i)|U^{-1}QU|\psi(\mathbf{r}_i)],$$

where we have used the definitions of Hermitian conjugate and unitary operators. Now if $\langle Q \rangle$ is *invariant* under the transformation induced by U then $\langle Q \rangle = \langle Q \rangle'$ so that

$$U^{-1}QU = Q,$$

but $\qquad U^{-1}U = 1,$

therefore $\quad U^{-1}QU = QU^{-1}U,$

that is Q commutes with U,

$$[Q, U] = 0.$$

In other words if an operator is invariant under a coordinate transformation induced by the unitary operator U, then these two operators commute.

3.3 Parity conservation

We have already made some study of the parity transformation. Here we may look at it from a somewhat different point of view, since it illustrates well the theory just described. If we write the Schrödinger equation as

$$H(\mathbf{r}_i)\psi(\mathbf{r}_i) = E\psi(\mathbf{r}_i),$$

where H is the Hamiltonian operator and E the energy eigenvalue, then it is clear that, for *isolated* systems, H and E do not change if we invert all the coordinates simultaneously through the origin; that is if we apply the parity transformation. Thus

$$H(\mathbf{r}_i) = H(-\mathbf{r}_i),$$

and replacing $\mathbf{r}_i$ by $-\mathbf{r}_i$ in the Schrödinger equation we get

$$H(\mathbf{r}_i)\psi(-\mathbf{r}_i) = E\psi(-\mathbf{r}_i),$$

so that $\psi(\mathbf{r}_i)$ and $\psi(-\mathbf{r}_i)$ satisfy the same equation. If there is no degeneracy (degeneracy may be removed formally by a small perturbation), then $\psi(\mathbf{r}_i)$ must be proportional to $\psi(-\mathbf{r}_i)$,

$$\psi(\mathbf{r}_i) = k\,\psi(-\mathbf{r}_i).$$

If we exchange $\mathbf{r}_i$ and $-\mathbf{r}_i$ we get

$$\psi(-\mathbf{r}_i) = k\,\psi(\mathbf{r}_i),$$

so that $k = \pm 1$.

We now see that the conservation of the parity of a system is a consequence of the equation of motion, and that it is true for systems obeying the Schrödinger equation. If we call the parity operator P and take a potential V which is symmetrical about the origin, so that the Hamiltonian H is invariant under the parity transformation, then P and H will commute. They thus have simultaneous eigenfunctions. The eigenvalues of P, ± 1, are also conserved, since we can write

$$i\hbar \frac{dP}{dt} = [P, H],$$

and if $[P, H] = 0$ then $dP/dt = 0$ and the eigenvalues of P are constants of the motion. Here we have the analogy to our earlier classical treatment where we wrote

$$\frac{d}{dt}\left(\frac{\partial L}{\partial \dot{q}_i}\right) = \frac{\partial L}{\partial q_i}.$$

For the classical case, the requirement for conservation of a quantity was that the Lagrangian be independent of that quantity. The parallel requirement in the quantum case is that the operator associated with a conserved observable must commute with the unitary operator which induces the transformation.

3.4 Operators and transformations in quantum theory

Often we are interested in transformations associated with the actual operators of quantum mechanics. First we require the connection between such operators (which are in fact Hermitian operators) and the unitary transformation operators; i.e. what unitary transformation is generated by a given Hermitian operator Q? These are related by the expression

$$U = e^{iQ} = 1 + iQ + \frac{(iQ)^2}{2!} + \frac{(iQ)^3}{3!} + \ldots.$$

Now consider the example of a transformation between two frames with only relative linear motion of constant velocity along the x-axis, so that the y- and z-coordinates are identical in both frames. The relativistic transformation equations (see Appendix A) then give for the coordinates in the two frames

$$x_1^2 - t_1^2 = x_2^2 - t_2^2 \quad \text{(writing } c = 1, \text{ as usual)},$$
$$y_1 = y_2,$$
$$z_1 = z_2.$$

We need a U such that $\psi' = U\psi$ or a Q such that

$$\psi' = \psi + iQ\psi + \frac{(iQ)^2}{2!}\psi + \ldots.$$

Consider frame 2 moved a small distance a relative to frame 1 from the moment when the origins are coincident, so that initially $x_1 = x_2$ and $t_1 = t_2 = 0$. Then, using Taylor's theorem, we have

$$\psi'(x, t) = U\psi(x, t) = \psi(x', t')$$

and $\quad \psi(x', t') = \psi(x, t) + a\dfrac{\partial}{\partial x}\psi(x, t) + t\dfrac{\partial}{\partial t}\psi(x, t) + \dfrac{a^2}{2!}\dfrac{\partial^2}{\partial x^2}\psi(x, t) + \ldots$

Suppose now that ψ is constant in time so that $\partial\psi/\partial t = 0$. Then

$$\psi(x', t') = \psi(x, t) + a\dfrac{\partial\psi}{\partial x} + \dfrac{a^2}{2!}\dfrac{\partial^2\psi}{\partial x^2} + \ldots$$

and we obtain $\quad U = 1 + a\dfrac{\partial}{\partial x} + \dfrac{a^2}{2!}\dfrac{\partial^2}{\partial x^2} + \ldots$

Now let a be a small unit distance along the x-axis so that we can write

$$U = 1 + \dfrac{\partial}{\partial x} + \dfrac{1}{2!}\dfrac{\partial^2}{\partial x^2} + \ldots = 1 + i\left(\dfrac{1}{i}\dfrac{\partial}{\partial x}\right) + \dfrac{i^2}{2!}\left(\dfrac{1}{i^2}\dfrac{\partial^2}{\partial x^2}\right) + \ldots$$

Thus we have

$$Q = \dfrac{1}{i}\dfrac{\partial}{\partial x},$$

corresponding, apart from the constant $\hbar$, to the linear momentum operator of quantum theory. Although we have considered only an infinitesimal transformation, we can perform a series of such transformations to achieve the finite transformation.

Consider now a displacement only along the t-axis. The derivative in question corresponding to Q is $i\hbar\ \partial/\partial t$, which we see to be proportional to the energy operator. We may recapitulate the steps in the argument linking symmetry under transformation and conservation laws as follows:

If we can show that $[Q, H] = 0$, then we can say that the eigenvalues of Q are constants, and we can see that the invariance of the eigenvalues of Q corresponds to symmetry under the transformation induced by Q. Thus, since the momentum operator commutes with H, the corresponding momentum eigenvalue is a constant, and the system is invariant under translation along the appropriate axis. The conservation of the energy eigenvalues corresponds to invariance under displacement in time.

3.5 **Angular-momentum conservation in quantum theory**

We have already discussed the conservation of angular momentum in the classical case. We may apply the principle of the previous section to the quantum-theory case. If we consider a rotation of the coordinate system by a small amount $\delta\phi$ about the z-axis then the wave function $\psi(x, y, z)$ becomes

$$\psi'(x, y, z) = \psi(x + \delta x, y + \delta y, z),$$

where $\delta x = y\ \delta\phi$ and $\delta y = -x\ \delta\phi$ and $\delta z = 0$. Using Taylor's theorem we can write

$$\psi'(x, y, z) = \psi(x, y, z) + y\,\delta\phi\,\frac{\partial\psi}{\partial x} - x\,\delta\phi\,\frac{\partial\psi}{\partial y} + \ldots$$

$$= \psi + \left(y\frac{\partial}{\partial x} - x\frac{\partial}{\partial y}\right)\psi\,\delta\phi + \ldots.$$

However we have that the angular momentum operator L_z is given by

$$-i\hbar\left(x\frac{\partial}{\partial y} - y\frac{\partial}{\partial x}\right)$$

and for an infinitesimal rotation we can neglect the higher-order terms and write

$$\psi'(x, y, z) \simeq \left(1 - \frac{i}{\hbar}L_z\,\delta\phi\right)\psi(x, y, z),$$

or, including the higher-order terms,

$$\psi'(x, y, z) = \psi(x, y, z) - \frac{i}{\hbar}L_z\,\delta\phi\,\psi + \frac{1}{2!}\cdot\left(-iL_z\frac{\delta\phi}{\hbar}\right)^2 + \ldots$$

$$= e^{-iL_z\,\delta\phi/\hbar}\,\psi(x, y, z).$$

Compounding many infinitesimal rotations to obtain the rotation ϕ we then obtain

$$\psi'(x, y, z) = e^{-iL_z\,\phi/\hbar}\,\psi(x, y, z).$$

The general expression for a rotation θ about a unit vector $\mathbf{n}$ is then

$$\psi'(x, y, z) = e^{-i\theta\mathbf{L}\cdot\mathbf{n}/\hbar}\,\psi(x, y, z).$$

Thus $e^{-i\theta\mathbf{L}\cdot\mathbf{n}/\hbar}$ corresponds to the unitary transformation operator and the operator $-\theta\mathbf{L}\cdot\mathbf{n}/\hbar$ commutes with operators associated with invariants. The energy is invariant under rotation in a system symmetrical about the axis of rotation so that

$$\left[-\frac{\theta\mathbf{L}\cdot\mathbf{n}}{\hbar}, H\right] = 0,$$

or, removing the numerical factor,

$$[L, H] = 0,$$

and the eigenvalues of L, i.e. the angular momenta, are conserved. Here we have neglected the effects of spin and also relativistic effects.

3.6 Conservation of isotopic spin

We shall deal here slightly more formally with this quantity which we have already introduced in Chapter 2.

The idea of isotopic spin was suggested for nucleons by Heisenberg shortly after the discovery of the neutron. Thus a nucleon is represented as a function of a dichotomic variable which can take one of two values, $f = \begin{bmatrix} a \\ b \end{bmatrix}$. The nature of such dichotomic functions is well known from the properties of particles of ordinary spin $\frac{1}{2}$. The standard linear operators which operate on such variables are the Pauli spin matrices and the unity operator. As usual it is convenient to use the half-values of the operators. These operators and their combinations are the only ones which can act upon a dichotomic function. Thus we have

$$2\tau_1 = \begin{bmatrix} 0 & 1 \\ 1 & 0 \end{bmatrix}, \quad 2\tau_2 = \begin{bmatrix} 0 & -i \\ i & 0 \end{bmatrix}, \quad 2\tau_3 = \begin{bmatrix} 1 & 0 \\ 0 & -1 \end{bmatrix}, \quad 2\tau_4 = \begin{bmatrix} 1 & 0 \\ 0 & 1 \end{bmatrix}.$$

We represent the proton as $\begin{bmatrix} 1 \\ 0 \end{bmatrix}$ and the neutron as $\begin{bmatrix} 0 \\ 1 \end{bmatrix}$. Then

$$2\tau_1 p = \begin{bmatrix} 0 & 1 \\ 1 & 0 \end{bmatrix}\begin{bmatrix} 1 \\ 0 \end{bmatrix} = \begin{bmatrix} 0 \\ 1 \end{bmatrix} = n$$

and similarly $2\tau_1 n = p$ so that τ_1 transforms $n \leftrightarrow p$, while the effects of τ_2 and τ_3 may be worked out as an example.

We may also examine the operators $(\tau_4 + \tau_3)$ and $(\tau_4 - \tau_3)$

$$(\tau_4 + \tau_3) = \begin{bmatrix} \frac{1}{2} & 0 \\ 0 & \frac{1}{2} \end{bmatrix} + \begin{bmatrix} \frac{1}{2} & 0 \\ 0 & -\frac{1}{2} \end{bmatrix} = \begin{bmatrix} 1 & 0 \\ 0 & 0 \end{bmatrix},$$

$$(\tau_4 + \tau_3)p = p \quad \text{and} \quad (\tau_4 + \tau_3)n = 0,$$

and similarly $(\tau_4 - \tau_3)p = 0$ and $(\tau_4 - \tau_3)n = n$.

Thus these operators will project out the proton or neutron parts respectively of a mixture of states. The step operators $(\tau_1 \pm i\tau_2)$ are also sometimes useful as in the angular-momentum case.

$$(\tau_1 \pm i\tau_2) = \begin{bmatrix} 0 & 1 \\ 0 & 0 \end{bmatrix} \text{ and } \begin{bmatrix} 0 & 0 \\ 1 & 0 \end{bmatrix}.$$

They have the properties

$$(\tau_1 + i\tau_2)n = \tau_+ n = \begin{bmatrix} 0 & 1 \\ 0 & 0 \end{bmatrix}\begin{bmatrix} 0 \\ 1 \end{bmatrix} = \begin{bmatrix} 1 \\ 0 \end{bmatrix} = p,$$

while $(\tau_1 - i\tau_2)p = \tau_- p = n$.

The τs of course obey the usual commutation rules for spin operators.

If we have more than one particle we use the method of product spaces so that, for instance, two particles are described in a four-dimensional space and n particles in a 2^n-dimensional space. In the two-particle case we may write vectors

$$pp = \begin{bmatrix} 1 \\ 0 \\ 0 \\ 0 \end{bmatrix}, \quad pn = \begin{bmatrix} 0 \\ 0 \\ 1 \\ 0 \end{bmatrix}, \quad np = \begin{bmatrix} 0 \\ 1 \\ 0 \\ 0 \end{bmatrix}, \quad nn = \begin{bmatrix} 0 \\ 0 \\ 0 \\ 1 \end{bmatrix}.$$

Since the i-spin for a nucleon is compounded of I_1, I_2 and I_3, which are proportional to the Pauli matrices, we can carry over all the normal rules of angular momentum to the i-spin 'space'. A properly constructed Hamiltonian involving forces which are *charge independent* is invariant under rotations in the i-spin space. Such a Hamiltonian then commutes with the corresponding rotation operators, and the i-spin is a constant of the motion. If the Coulomb force, for say a pair of protons, is included in the Hamiltonian then H no longer commutes with the total i-spin. Here we have a clear example of an invariance principle which is true for the 'strong' interactions but not for the electromagnetic interactions. We have already examined the relationship between i-spin and charge. We can also treat this in terms of a charge operator, the eigenvalues of which for the nucleon system must be $+1$ and 0. We write

$$Q = (\tfrac{1}{2}+I_3)$$

so that $Qp = +1p$ and $Qn = 0n$.

For two nucleons we have $Q = Q_1+Q_2 = 1+(I_3^1+I_3^2)$. For N nucleons

$$Q = \tfrac{1}{2}N+\sum_N I_3$$
$$= \tfrac{1}{2}N+T_3,$$

where we have written T_3 for the sum of the i-spin third components. Since Q and N have conserved eigenvalues (N here is simply a 'number operator' for nucleons) so must T_3. The eigenvalues of T_3 are then $-T, -T+1, \ldots, +T$, where T is the eigenvalue of the total i-spin operator I^2, i.e.

$$I^2\psi = T(T + 1)\psi.$$

Thus the operator Q has eigenvalue q where

$$q = -T+\tfrac{1}{2}N, -T+1+\tfrac{1}{2}N, \ldots, +T+\tfrac{1}{2}N.$$

The centre of charge of this isotopic-spin multiplet is $\tfrac{1}{2}N$. In fact we shall see that it is always true for non-strange particles that

Centre of charge $= \tfrac{1}{2}$(baryon number),

where the baryon number in the present case is simply the total number of nucleons, and will be discussed further in section 3.8.

For $I = 1$ we have an i-spin triplet with $I_3 = -1, 0, +1$. In nuclei this triplet corresponds to the three neighbouring isobars having charges $\tfrac{1}{2}N-1, \tfrac{1}{2}N, \tfrac{1}{2}N+1$. For $I = \tfrac{1}{2}$ we have isotopic nuclear doublets with charges $\tfrac{1}{2}N-\tfrac{1}{2}$ and $\tfrac{1}{2}N+\tfrac{1}{2}$. Such an i-spin doublet is the pair ${}^7_3\text{Li}$, ${}^7_4\text{Be}$,

${}^7_3\text{Li}$ 3 protons $q = 3,$ $T = \tfrac{1}{2},$ $T_3 = -\tfrac{1}{2},$

${}^7_4\text{Be}$ 4 protons $q = 4,$ $T = \tfrac{1}{2},$ $T_3 = +\tfrac{1}{2}.$

In accord with charge independence, the ground and first excited states of this doublet are indeed found to be closely similar when allowance is made for the Coulomb effects.

We have already seen that the existence of three charged states for the pion implies an i-spin of 1 for this particle. This is in accord with the requirement from the basic Yukawa process

$$N \rightarrow N + \pi \quad \text{(nucleon} \rightarrow \text{nucleon} + \text{pion)},$$

in which i-spin conservation would limit the pion i-spin to 0 or 1. We have noted that here also the relationship between charge I_3 and baryon number is

$$Q = I_3 + \tfrac{1}{2}N,$$

and that the centre of charge is again given by $\tfrac{1}{2}N$, since for pions $N = 0$.

The i-spin operators for pions are then the same as those for a spin-1 system

$$\rho_1 = \frac{1}{\sqrt{2}} \begin{bmatrix} 0 & 1 & 0 \\ 1 & 0 & 1 \\ 0 & 1 & 0 \end{bmatrix}, \quad \rho_2 = \frac{1}{\sqrt{2}} \begin{bmatrix} 0 & -i & 0 \\ i & 0 & -i \\ 0 & i & 0 \end{bmatrix}, \quad \rho_3 = \begin{bmatrix} 1 & 0 & 0 \\ 0 & 0 & 0 \\ 0 & 0 & -1 \end{bmatrix},$$

while the i-spin wave functions are

$$\pi^+ = \begin{bmatrix} 1 \\ 0 \\ 0 \end{bmatrix}, \quad \pi^0 = \begin{bmatrix} 0 \\ 1 \\ 0 \end{bmatrix}, \quad \pi^- = \begin{bmatrix} 0 \\ 0 \\ 1 \end{bmatrix}.$$

It is easy to check the effects of these operators, e.g.

$$\rho_1 \pi^+ = \frac{1}{\sqrt{2}} \pi^0, \quad \rho_1 \pi^0 = \frac{1}{\sqrt{2}} (\pi^+ + \pi^-), \quad \rho_1 \pi^- = \frac{1}{\sqrt{2}} \pi^0.$$

We have already considered the problem of dealing with systems consisting of pions and nucleons which is exactly analogous to the combination of angular momenta $\tfrac{1}{2}$ and 1 and we shall not develop this formalism further here.

3.7 The generalized Pauli principle

We have now studied the ideas which are necessary to make a generalization of the Pauli exclusion principle. We recall that Pauli proposed that the electron structure in atoms was consistent with the hypothesis that no quantum state could be occupied by more than one electron. This result follows from certain anti-commutation relations among the operators for the Dirac field appropriate for spin-$\tfrac{1}{2}$ particles, and it is beyond the scope of the present discussion to deal with this aspect more deeply. However we can show that the Pauli principle as stated above implies that only *antisymmetric* overall wave functions are possible for the electron or other *fermion* systems.

Formally this result can be seen to arise in the following way. We write a wave function for two identical particles in terms of quantum numbers for each particle, which we group as a vector **x**. Thus for particles 1 and 2 we write

$$\psi(\mathbf{x}_1, \mathbf{x}_2).$$

Since the physical situation described by the wave function must be the same when the identical particles are interchanged we have

$$|\psi(\mathbf{x}_1, \mathbf{x}_2)|^2 = |\psi(\mathbf{x}_2, \mathbf{x}_1)|^2$$

so that $\psi(\mathbf{x}_1, \mathbf{x}_2) = \pm e^{i\phi} \, \psi(\mathbf{x}_2, \mathbf{x}_1),$ **3.1**

and we write the unobservable phase factor equal to one, since if we repeat the interchange of particles we retrieve the original state and the phase factor becomes $e^{2i\phi} = 1$ so that $e^{i\phi} = \pm 1.$

We now write $\psi(\mathbf{x}_1, \mathbf{x}_2)$ in terms of a linear superposition of product wave functions $\psi_1(\mathbf{x}_1) \, \psi_2(\mathbf{x}_2)$. Including a normalization factor 2 we can write

$$\psi(\mathbf{x}_1, \mathbf{x}_2) = \frac{\psi_1\psi_2 + \psi_2\psi_1}{\sqrt{2}}$$ **3.2**

and $\psi(\mathbf{x}_1, \mathbf{x}_2) = \dfrac{\psi_1\psi_2 - \psi_2\psi_1}{\sqrt{2}},$ **3.3**

both of which satisfy the relation **3.1**, but the first of which is *symmetric*, since it does not change sign on interchange of particles 1 and 2, and the second of which is *antisymmetric*. However if $\psi_1 \equiv \psi_2$ the expression **3.3** vanishes while **3.2** does not, so that satisfaction of the Pauli principle is seen to require a wave function which is antisymmetric. The argument may be extended to show that, for any number of identical particles, the wave function must be antisymmetric under interchange of any two.

For bosons, on the other hand, the total wave function is *symmetric* and the exclusion principle as such no longer holds.

These results concerning symmetry have some interesting consequences for the possible states of systems consisting of bosons or fermions. We consider a system consisting of two identical fermions of spin $\frac{1}{2}$. The wave function for such a system may be written as the product of a space-dependent part $\xi(\mathbf{r}_1, \mathbf{r}_2)$ and a spin-dependent part $\chi(S, m_1, m_2)$ where S is the total spin and m_1 and m_2 are the individual z-components, $\pm\frac{1}{2}$. Clearly parallel-spin states ($S = 1$) are symmetric, while antiparallel-spin states ($S = 0$) are antisymmetric, as can be seen by writing **3.2** and **3.3** for the spin wave function.

$$\chi = \chi_1\chi_2 = \frac{\chi_1\chi_2 \pm \chi_2\chi_1}{\sqrt{2}}.$$

The orbital angular-momentum wave function we have already seen to be symmetric or antisymmetric according as the orbital angular-momentum quantum number l is even or odd (section 2.7). Thus we may summarize the behaviour of the overall wave function under exchange of the two particles as

$$(-1)^l (-1)^{S+1}$$

and according to the Pauli principle we must make this antisymmetric, i.e. equal to -1 so that $(l+S)$ must be even. Thus parallel-spin states can exist only for odd orbital angular momentum, and vice versa. In the usual spectroscopic notation this means that permissible states for two spin-$\frac{1}{2}$ particles are

$$^1s_0, \, ^3p_{0,\,1,\,2}, \, ^1d_2, \, \ldots .$$

For spinless bosons on the other hand l must be even.

We may readily extend the wave function and the Pauli principle to include the isotopic spin. As with the ordinary spin we can write an i-spin wave function $\phi(I, I_{1z}, I_{2z})$. The Pauli principle then yields

$$(-1)^l (-1)^{S+1} (-1)^{I+1} = -1,$$

so that $(l+S+I)$ must be odd.

One consequence of this rule is that a proton–neutron pair in an $I = 1$ state must have even $(l+S)$ and in an $I = 0$ state odd $(l+S)$. The application to nuclear- and i-spin is discussed in Gibson (1971).

3.8 Additive conservation laws

A somewhat different kind of conservation law from those discussed in the preceding sections of this chapter is the so-called 'additive' conservation law which requires that the algebraic sum of certain quantum numbers remains constant throughout all possible processes, or throughout interactions of a particular kind. The most familiar such quantum number, which is conserved in all known interactions, is electric charge. This is sometimes known as an 'internal' quantum number since it cannot be visualized as having any relation to an external coordinate system. As we shall see, there are other such internal quantum numbers.

In view of the relationship between symmetry under transformation and conservation laws which we have discussed already, it is natural to ask what transformations are involved for the additive conservation laws. The invariance is under the so-called gauge transformations of the first kind.

For such a transformation the wave functions of all the particles in the state in question are multiplied by a phase factor:

For the ith particle

$$\psi_i(x, y, z, t) \rightarrow e^{i\alpha} \, \psi_i(x, y, z, t),$$

where α is a real constant.

For such a transformation of the wave functions we require corresponding transformations of the field amplitudes (recall the Yukawa equation of Chapter 2 involving the field amplitude ϕ). The transformations are as follows:

$$\phi_i(x, y, z, t) \rightarrow e^{i\alpha A} \, \phi_i(x, y, z, t),$$

where A is a real number. It may be shown that invariance of the Lagrangian under such transformations leads to a continuity relationship for the particle and 'current' densities. The continuity equation implies conservation of the algebraic sum of the appropriate quantum numbers.

3.9 Antiprotons and the conservation of baryons

An antiparticle to the electron (and thus most probably to the proton and neutron) was predicted by Dirac in 1930, as a consequence of his relativistic wave equation. Particle and antiparticle were predicted to annihilate into quanta of the interaction field. The antiparticle of the electron was observed by Anderson in a cloud chamber in 1932, during studies of the cosmic radiation, as a *positively charged* electron or *positron*. Subsequent studies of positrons have demonstrated that they annihilate with electrons into γ-rays.

The antiproton, according to the argument to be presented below, can only be produced in proton–nucleon collisions at energies greater than 5·6 GeV, while in a nuclear target the threshold is 4·3 GeV due to the Fermi motion. Thus only when the bevatron accelerator with an accelerated proton energy of 6 GeV came into operation could antiproton production in the laboratory be expected. The anti-protons were sought and found in an experiment carried out in 1955 by Chamberlain, Segrè, Wiegand and Ypsilantis. The fundamental problem is to sort

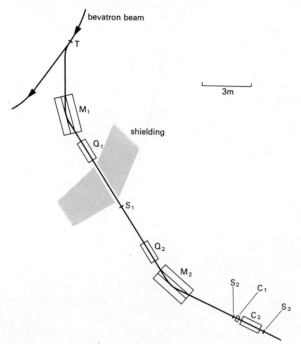

Figure 32 The experimental arrangement used by Chamberlain, Segrè, Wiegand and Ypsilantis (1955) in the first observation of antiprotons.
M_1 and M_2 are bending magnets, Q_1 and Q_2 quadrupole focusing magnets, S_1, S_2 and S_3 plastic scintillation counters, C_1 a Cerenkov counter containing $C_8F_{16}O$ ($\mu = 1·276$) and C_2 a Cerenkov counter of fused quartz ($\mu = 1·458$)

out the rather small number of antiprotons produced at this energy from the very much greater number of negative mesons produced in the Bevatron target, a background which exceeds the number of antiprotons by a factor of about 10^5.

The experimental arrangement is shown in Figure 32. Only negative particles originating at the Bevatron target and having a momentum such that the magnets M_1 and M_2 guide them through the system are counted. The selected momentum was $1 \cdot 19 \, \text{GeV}/c$. At this momentum pions have a velocity of $0 \cdot 99 \, c$, kaons of $0 \cdot 93 \, c$, and protons of $0 \cdot 78 \, c$. The particles were then distinguished by a measurement of their velocity by means of the scintillation counters S_1 and S_2 placed at beam foci produced by the quadrupoles Q_1 and Q_2. For protons the time to traverse the 12 m between S_1 and S_2 was 51 ns (1 ns $= 10^{-9}$ s), while for kaons it was 43 ns. Thus the S_1 signal was delayed by 51 ns, and coincidences sought between the delayed S_1 signal and the pulse from S_2. With a background of magnitude such as

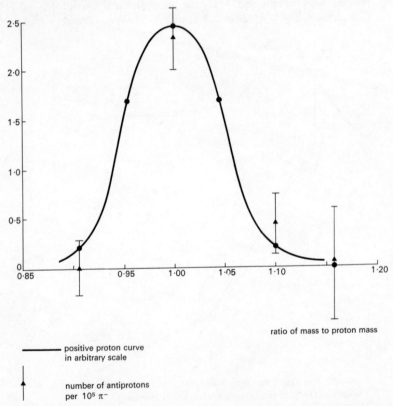

ratio of mass to proton mass

—————— positive proton curve
in arbitrary scale

number of antiprotons
per $10^5 \, \pi^-$

Figure 33 The counting rate in the antiproton experiment as a function of particle mass, as determined by changing the momentum selection and keeping the velocity selection unchanged. The dot points were determined using protons. The antiproton points are seen to lie on the same curve

in this experiment additional means of discrimination are desirable, and were provided by the Cerenkov counters C_1 and C_2; C_1 is sensitive to all charged particles with velocity greater than $0.79\ c$. By the time an antiproton had passed through S_2, C_1 and C_2 its velocity was reduced to $0.765\ c$ and C_2 was a counter designed to count particles with velocity between $0.75\ c$ and $0.78\ c$. The signal for an antiproton is thus counts in S_1, S_2, S_3 and C_2 with no count in C_1 and an S_1–S_2 delay of 51 ns. This system eliminates practically all forms of background. The result obtained is illustrated in Figure 33, where the solid curve gives the mass resolution as obtained by running protons through the system, while the triangles indicate the antiproton measurements.

In the years since this initial observation, the advent of higher-energy accelerators and the development of particle-beam technology have rendered the production of secondary *beams* of antiprotons standard practice in high-energy laboratories.

We have already met the baryon number in our discussion of isotopic spin where we saw that the charge was related to the third component of the i-spin by $Q = I_3 + \frac{1}{2}N$, where for nucleons $N = 1$ and mesons $N = 0$. Before the recent discovery of some very heavy meson resonances, it was possible to define a baryon as a strongly interacting particle of mass equal to, or greater than, that of the nucleons, and which cannot decay totally into leptons (see next section) as can mesons. However it is now known that there exist mesons with masses greater than the nucleon mass, and this definition is no longer adequate. If we know the i-spin for a multiplet of particles we can obtain the baryon number from the above relationship, otherwise we depend on the conservation law for baryon number, which is experimentally shown to be true for nucleons and mesons and which is confirmed by, or found to be consistent with, all reactions involving the known particles. The conservation law operates as follows: baryon number $+1$ is assigned to the nucleons and -1 to the antinucleons. We then see that the conservation of the baryon number will allow some processes but forbid others. For instance, consider the production of antiprotons in proton–proton collisions. If baryon number need not be conserved, then antiprotons could be produced in the reaction

$$p + p \rightarrow p + \bar{p} + \pi^+ + \pi^+,$$

the threshold for which is only 600 MeV. However, when baryon number conservation is required, the cheapest way of producing antiprotons is by means of the process

$$p + p \rightarrow p + p + p + \bar{p},$$

for which the threshold is 5.6 GeV kinetic energy for the incident proton. The fact that antiprotons are experimentally observed to be produced in the second reaction, but not in the first, is a confirmation of the baryon conservation law. Other examples may be quoted, such as in antiproton annihilation we have that

$\bar{p} + p \rightarrow \bar{n} + n$ is allowed,
$\bar{p} + p \rightarrow n + n$ is forbidden,
$\bar{p} + p \rightarrow \pi^+ + \pi^-$ is allowed,
$\bar{p} + p \rightarrow p + \pi^-$ is forbidden.

We see that the assignment of baryon number -1 to the antinucleons is consistent with the charge–I_3 relationship. If we investigate the corresponding invariance to this conservation law, we find that it implies invariance under a gauge transformation of the field variables

$$\phi_i \rightarrow e^{i\alpha B_i} \phi_i,$$

where B_i is the baryon number for the ith particle, and B has been used rather than N to emphasize that the hyperons (see Chapter 5) are also baryons.

3.10 Lepton conservation

The leptons may be defined as those particles specifically associated with 'weak' interactions (see Chapter 4). The known leptons are also those particles with masses less than the π-meson mass; i.e. the electron, the muon, the neutrino and their corresponding antiparticles. As with the baryons, if we assign a 'lepton number' which is $+1$ for particles and -1 for the antiparticles (it does not matter which convention one adopts), then we find that this number is conserved for any system. We may state the law alternatively by saying that the difference between the number of leptons and the number of antileptons in any weak interaction is a constant. The corresponding gauge transformation is then

$$\phi_i \rightarrow e^{i\alpha l_i} \phi_i,$$

where l_i is the lepton number for the ith particle.

We may assign lepton numbers, in the first instance, by examining some of the observed processes, and we may then examine whether these assignments are consistent with all the processes observed and apparently forbidden. If we take the usual β-decay process

$$n \rightarrow p + e^- + \bar{\nu},$$

then the electron and the antineutrino must have opposite lepton numbers. Similarly for the decay

$$\pi^- \rightarrow \mu^- + \bar{\nu},$$

we see that the μ^- and the antineutrino have opposite lepton numbers. Let us assign lepton number $+1$ to the e^-, μ^-, and ν, and lepton number -1 to their antiparticles e^+, μ^+, and $\bar{\nu}$. Then we see that in the muon decay

$$\mu^+ \rightarrow e^+ + \nu + \bar{\nu}$$

the two neutrinos must be one antineutrino and one normal neutrino. We also note that the muon cannot decay into two leptons with conservation of lepton number.

It is clear that in fact the baryon and lepton numbers behave in exactly the same way as the electric charge of a particle. They are intrinsic properties of the particle and the total baryonic, leptonic and electric charges of any system are conserved in all processes.

3.11 Muon and electron conservation, and the two neutrinos

As we shall see below it has been conclusively shown that there exist two distinct varieties of neutrino, one of which is always associated with electrons, and the other always with muons. We may thus make the conservation law for leptons even more specific, and state that the total 'electron' number and the total 'muon' number are separately conserved. By electron, in this context, we mean electron, positron, electron neutrino and electron antineutrino, and by muon we mean positive and negative μ-meson with the corresponding muon neutrino and anti-neutrino. The difference between these two neutrinos, and the corresponding separate muonic and electron number conservation laws, can only be established by means of a study of neutrino absorption, and not from decay processes; in such processes we have no means of checking the nature of the neutrino produced.

An experiment which conclusively established the existence of two varieties of the neutrino was performed at the Brookhaven Laboratory proton synchroton in 1962. The principle of this experiment was to establish whether neutrinos arising from the decay

$$\pi^+ \rightarrow \mu^+ + \nu$$

would, when absorbed by protons or neutrons, give rise only to muons or to muons and electrons according to the reactions

$$\bar{\nu} + p \rightarrow n + e^+,$$

$$\bar{\nu} + p \rightarrow n + \mu^+.$$

The cross-section for the absorption reactions was expected to be very small, of the order of 10^{-36} mm^2, on the basis of weak-interaction calculations. This cross-section is for neutrinos having an energy of about one to two thousand million electronvolts. At lower energies the cross-sections are even smaller since the phase-space factor is smaller. Thus in order to study these processes one requires a very high flux of neutrinos having a high energy. The advent of the very-high-energy accelerators at Brookhaven and at the European Laboratory CERN, at Geneva, made it possible to produce beams of neutrinos satisfying these conditions. The arrangement used in the original experiment at Brookhaven was one in which the circulating proton beam in the proton synchrotron, accelerated to an energy of 15 GeV and having an intensity of about 2×10^{11} protons per second, was allowed to strike a target of beryllium in a three-metre straight section. An intense flux of pions (as well as other particles) was emitted from the target, predominantly in the forward direction. The pions were allowed to decay along a twenty-one-metre flight path, and the muon and neutrinos from this decay were also collimated in the forward direction due to the centre-of-mass motion. About 10 per cent of the pions decayed in the flight path. At the end of the flight path all particles except the neutrinos were stopped by a wall consisting of thirteen metres of solid iron. The neutrinos, being so weakly interacting, were largely unaffected by this material and passed into an arrangement of large spark chambers, with their lines of flight undeviated from the original direction. The spark chambers used in the Brookhaven experiment had dimensions $1\frac{1}{3}$ m by $1\frac{1}{3}$ m by $\frac{1}{3}$ m in thickness, and were each composed of nine 25 mm thick aluminium plates. The arrangement is shown schematically in Figure 34. The total mass of material in the chambers was about ten thousand kilogrammes.

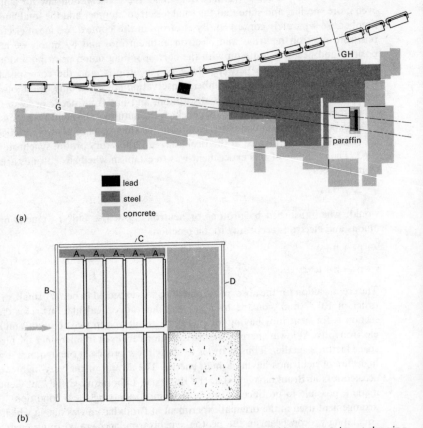

(a)

(b)

Figure 34 (a) Plan view of the arrangement for the Brookhaven experiment showing part of the accelerator, the shielding and position of the spark-chamber arrangement. (b) The spark-chamber counter arrangement in the Brookhaven neutrino experiment. A are the triggering counters, B, C and D are anticoincidence counters (Danby et al., 1962)

Since the cross-section for neutrino interactions is so small, it is essential to avoid even very weak backgrounds arising from cosmic rays or other sources. This was achieved by two methods. (a) The spark chambers were surrounded by scintillation counters in anticoincidence with the triggering of the scintillation counters placed *between* the spark chambers. This arrangement eliminated to a large extent the cosmic-ray background and also any very energetic muon penetrating the shield. The requirement for triggering was a count in any two of the internal counters, with no count in the surrounding anticoincidence counters. (b) The chambers were switched on only for the very brief period when particles were actually available from the target of the synchroton. In fact the beam pulse lasted only 25 μs, and within this there was a structure consisting of twelve bunches, each of which was about 20 ns long, separated by 220 ns. The chambers were then gated by means of a signal from a Cerenkov counter looking at the synchrotron target, and were opened for only 3·5 μs per pulse. This resulted in a most remarkable feature of the experiment in which, although 1 700 000 pulses were accepted during the experiment, the total sensitive time of the chamber was only $5\frac{1}{2}$ s, thus effecting a most useful limitation of cosmic-ray background.

The results of the experiment were the observation of over a hundred events in the spark chamber. The problem was then to determine (a) whether these events are in fact due to genuine neutrino absorptions, and if so (b) whether the absorptions resulted in the production of muons and of electrons, or of muons only. We first consider possible production of events by cosmic rays. The apparatus was operated for a period while the accelerator was not running, in order to give a measure of the expected number of cosmic-ray background events. This indicated that about five events were to be expected in which a long track, having a momentum greater than 300 MeV/c if it were a muon, was produced. In fact thirty-four such events were observed in the experiment. Secondly, it was necessary to test the possibility that the events might be due to neutrons which had somehow penetrated the shield. In fact in the early part of the run there was some indication of such events, originating in a certain section of the chamber, due to neutrons which had leaked past the shield; they were eliminated in the second section of the run by the addition of extra shielding. These gave rise to short-track events. For the long tracks, and an additional twenty-two events showing a vertex, the evidence against production by neutrons is (a) there was no attenuation in the production of events along the length of the chamber and (b) the removal of 1·2 m of iron from the main shielding did not produce any observable increase in the event rate, where a substantial increase would have been expected for neutrons in the beam direction. That the events were produced by pion- (or kaon-, see later) decay products was established by removing 1·2 m of iron from the shielding wall and replacing it by lead at a distance of only $1\frac{1}{2}$ m from the target thus maintaining the same total interaction path but reducing the flight path of the pions so that 90 per cent of them were eliminated by interaction before they had time to decay. The event rate was found to be reduced to $0·3 \pm 0·2$ events per 10^{16} protons, compared with the previous number of $1·46 \pm 0·2$ events per 10^{16} protons, quite consistent with the expected numbers if the events were due to the neutrinos from pion decay.

Having established that the events were indeed due to neutrino absorptions, the problem was to identify the secondary tracks. If the secondary tracks were muons they should not interact in the spark-chamber plates. In the second half of the run, where the background had been minimized, 8·2 m of aluminium was traversed by single tracks, no case of nuclear interaction being observed. Had these tracks been π-mesons a total of eight nuclear interactions would have been expected on the basis of calibration experiments. In addition, had the charged tracks been π-mesons one would have expected also about fifteen neutral pions. No such neutral pions were observed. The chamber sensitivity to electron showers was calibrated by putting electrons of various energies into the chamber. These produced fairly characteristic short tracks with discontinuities. In the clean part of the run no such showers were observed.

Thus it seems clear that the absorption of these muon-associated neutrinos gives rise predominantly, and probably exclusively, to production of muons, while if there is only one variety of neutrino we might have expected equal numbers of muons and electrons.

Both the theoretical interpretation and the quantity and quality of the experimental data have been improved since the first experiment. Detailed consideration of the possible interactions for the absorption mechanism has produced no theory which is consistent with the experimental data and would allow an explanation in terms of only one type of neutrino. Following the Brookhaven experiment a more elaborate experiment was performed at CERN in which the flux of neutrinos achieved was very much improved, due particularly to the use of an extracted proton beam from the accelerator. This beam was allowed to produce pions in a target placed within a special magnetic-focusing device (the magnetic 'horn') which concentrated into the forward direction pions having a wide range of momenta and angles of emission. In addition, the detectors used in this work were more sophisticated, consisting of a spark-chamber arrangement in which the tracks were bent in a magnetic field, facilitating momentum measurement, and in addition a large heavy-liquid bubble chamber. Although a larger number of events was observed in the spark chamber, those observed in the bubble chamber allowed a very complete analysis. The data obtained confirmed the original two-neutrino result and allowed a more detailed study of the neutrino-absorption process.

Chapter 4
Strong, weak and electromagnetic interactions

4.1 **Types of interactions**

In Chapter 2 we have seen that the nuclear interaction is very much stronger than the electromagnetic interaction, their relative strengths being characterized by the difference in the 'coupling constants'. We may recognize at this stage two other forms of interaction, gravitational forces and the so-called 'weak' interaction. Qualitatively we distinguish these interactions by both their different strengths or coupling constants and by the nature of the particles taking part in them. We may summarize as follows:

Electromagnetic interactions give rise to the emission and absorption of real or virtual photons by charged particles.

Strong interactions are responsible for the interaction between nucleons and in particular for the Yukawa process

$$N \leftrightarrow N + \pi,$$

so that they are apparently mediated by mesons. We shall see later that there exist strong interactions between particles other than the nucleons.

Weak interactions which are responsible for many characteristic lepton phenomena such as ordinary β-decay and pion- and muon-decay.

Gravitational interactions which exist between all particles having mass, and which are familiar and important on the macroscopic scale but unimportant in discussing elementary-particle phenomena.

We note that the quanta of the electromagnetic and strong interactions have been identified as the photon and the meson (the π-mesons which we have already discussed and heavier mesons which will be dealt with later), while the quanta of the weak and gravitational processes have not been detected. Much effort has been put into the search for such quanta, and it is appropriate here to discuss briefly the present position with regard to the proposed quantum of the weak-interaction field, the so-called intermediate vector boson.

On the same basis on which one may look for the production of free π-mesons in the strong interactions of nucleon collisions, if the energy is sufficiently high to supply the meson mass, one might expect to be able to produce the intermediate vector boson, commonly denoted by the letter W, in collisions of high-energy neutrinos with nucleons. Thus we compare the processes shown in Figure 35.

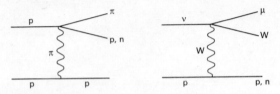

Figure 35

In order to produce the W we might then use the reactions

$$\nu_\mu + p \rightarrow p + \mu^- + W^+, \qquad\qquad\textbf{4.1}$$

$$\bar{\nu}_\mu + p \rightarrow p + \mu^+ + W^-. \qquad\qquad\textbf{4.2}$$

We now consider what may be the decay modes to be expected for the W-particle: it cannot be stable, otherwise it would surely have been detected. In fact its lifetime must be rather short, since it has never been observed in experiments, and one would expect an upper limit of about 10^{-16} s. The decay modes possible will of course depend on the mass of the W-particle which we have no means of estimating in advance. For decay modes into leptons we should certainly expect the processes

$$W \rightarrow e + \nu,$$

$$W \rightarrow \mu + \nu,$$

while it is also possible that it should decay into pions. It was naturally a prime objective of the various neutrino experiments to search for the existence of the intermediate vector boson. If we refer to the possible production processes **4.1**, **4.2**, then for the muon decay of the W we would expect to see neutrino absorption events which gave rise to a muon pair. A detailed study of all possible muon pairs, in particular of the ranges of the positive and negative particles, yields the result that the observations do not agree with the expectations for the W-decay. For the neutrino energies used the data thus indicates that the W is not produced, possibly because its mass is too great. The lower limit on the mass of the W which can be set by these experiments is about $1 \cdot 8$ GeV/c^2.

4.2 Coupling constants: dimensions

In order to compare the nature of observed processes we frequently wish to determine the coupling constants involved. The processes in question are either reactions between bombarding and target particles, for which we can measure the cross-section, or the decay of particles, for which we can measure the lifetime. We shall treat these processes in more detail below, but it is instructive first to examine the dimensions to be expected for the coupling constants. A simple and useful description of processes for this purpose was given by Jauch (1959).

We note first that in considering interactions it is necessary to distinguish between bosons and fermions, since bosons can be absorbed or emitted singly but fermions only in pairs. If we then write the symbol ϕ to indicate boson emission or absorption and the symbols ψ, $\bar{\psi}$ to indicate fermion emission or absorption, we can represent any process in terms of these quantities. Thus, for instance, for a process involving one boson and two fermions, such as the decay

$$\pi \rightarrow \mu + \nu,$$

we may draw a diagram as in Figure 36.

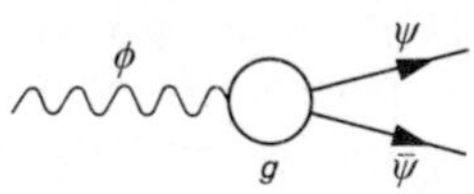

Figure 36

g is the coupling constant measuring the interaction strength, and we can write a matrix element for the process as $g\phi\psi\bar{\psi}$. The notation is useful either if we know the actual forms of ϕ and ψ or in *comparing* reactions, even where the actual forms are not known.

We can obtain the dimensions of g if we write the general term for an interaction as

$$g\phi^B(\psi\bar{\psi})^F$$

(where B is the number of bosons and F the number of fermion pairs) and examine the dimensions of ϕ and ψ. Using units such that $\hbar = c = 1$, leaving only the free dimension L then we have for mass the dimensions:

(joule seconds)(velocity)$^{-2}$(time)$^{-1}$: $(\hbar)(c^2L)^{-1} = L^{-1}$.

Similarly, both momentum and energy have dimensions L^{-1}. We now need two results from field theory, namely that the energy density in a boson field, due to the self-energy of a particle of mass m, is $(m^2\phi^2)$ and correspondingly in a fermion field it is $(m\psi\bar{\psi})$. In our system of units, energy density has the dimensions of L^{-4}, so that ϕ^2 has dimension L^{-2}, and ϕ dimension L^{-1}, while $\psi\bar{\psi}$ has dimension L^{-3}, and ψ has dimension $L^{-\frac{3}{2}}$. Then the matrix element, or Hamiltonian density, has dimension L^{-4}. Using the above results we can write

$$g\phi^B(\psi\bar{\psi})^F = gL^{-B-3F} = L^{-4},$$

so that g has dimensions $L^{-4+B+3F}$. Thus, for the pion-decay interaction, g has dimension L^0. The electromagnetic interaction is also like this; for instance pair production corresponds to the same diagram, where ϕ represents the photon, ψ and $\bar{\psi}$ the electron–positron pair, and where g^2 is again dimensionless, being given by $\frac{1}{137}$. The Yukawa process $N \leftrightarrow N + \pi$ is also of this form. The only other dimensionless coupling constant is ϕ^4.

4.3 Coupling constants and decay lifetimes

The problem we shall consider in this section is how to calculate, from the measured value of the decay lifetime, the coupling constant effective in a decay process. We start from the usual expression for a transition rate

$$R = 2\pi |M|^2 \frac{dN}{dW},$$

where we have used units $\hbar = c = 1$, $|M|^2$ is the square of the modulus of the appropriate matrix element and where dN/dW is the density of final states. For interactions of the type involving one boson and two fermions ($g\phi\psi\bar{\psi}$) it can be shown that M, the non-Lorentz invariant matrix element, (see Appendix A) may be written as

$$\frac{gSV^{-\frac{1}{2}}}{\sqrt{(2e_B)}},$$

where e_B is the boson energy and S is a factor depending upon the spins of the particles involved. This spin factor we shall set equal to unity for the purpose of the present calculation, which is intended to obtain the order of magnitude of g rather than any precise value. V is a normalization volume and the factor $V^{-\frac{1}{2}}$ arises since we must ensure that

$$\int \psi^* \psi \, d\tau = 1.$$

We can see that the factor $1/\sqrt{(2e_B)}$ is necessary, for dimensional reasons, in view of our arguments concerning the dimensions of the coupling constants. It can be justified properly only by deeper analysis than can be given here.

Using the above result we then obtain the probability of for instance the decay $\Lambda \rightarrow p + \pi$

$$\frac{g^2 \pi}{V e_\pi} \frac{dN}{dW},$$

or for the lifetime

$$\tau = \frac{e_\pi V}{g^2 \pi} \left(\frac{dN}{dW} \right)^{-1}. \tag{4.3}$$

We require an expression for the phase-space factor, which for the two-particle decay can be written (see Appendix A)

$$\frac{dN}{dW} = \frac{V p^2}{2\pi^2 (v_1 + v_2)},$$

where p is the c.m.s. momentum of the decay products, and v_1 and v_2 are the velocities of the products, also in the c.m.s. If we write M_0 as the mass of the decaying particle, and e_1 and e_2 as the energies of the decay products in the c.m.s., then we obtain for the phase-space factor

$$\frac{dN}{dW} = \frac{Vpe_1e_2}{2\pi^2 M_0}.$$

When substituted into the equation **4.3** this yields an expression for the lifetime

$$\tau = \frac{4\pi}{g^2} \frac{M_0}{2pe_p},$$

where e_p is the decay-proton energy. This is sometimes written in units of $\tau_0 = \hbar/m_e c^2 \simeq 10^{-21}$ s, giving the result

$$\tau = \tau_0 \frac{4\pi m_e M_0^2}{g^2 M_0 2pe_p}.$$

This relationship enables us to calculate the value of the coupling constant from the measured value for the lifetime. If we take the observed value for the lifetime of the lambda decay $\Lambda \to p\pi$ to be $2\cdot5 \times 10^{-10}$ s, then we obtain a value for $g^2/4\pi$ of $1\cdot5 \times 10^{-14}$. As we shall see later, very similar values are obtained for a number of other decays involving one boson and two fermions.

4.4 Coupling constants and reaction cross-sections

For the Yukawa-type process, which also has the form $\phi\psi\bar{\psi}$, we have no naturally occurring decays, since none are allowed by energy conservation. For such interactions we must study the coupling constant by measuring reaction cross-sections. However, in such situations we have at least two particles in the final state as well as two in the initial state. Thus the form of the interaction for processes such as pion–proton scattering is $\phi^2\psi\bar{\psi}$, since we have two bosons and two fermions. The dimensions of the coupling constant are then seen to be L^{+1}. The matrix element now has the form

$$M = \frac{g}{(2e_{B_1})^{\frac{1}{2}}(2e_{B_2})^{\frac{1}{2}}V},$$

where e_{B_1} and e_{B_2} are the boson energies. Again the energy factors can be rigorously obtained only from field theory, although as before we can see that such factors are necessary for dimensional reasons. V represents the normalization volume. In this case what we can measure is the value of the cross-section, so that we must derive a relationship between the cross-section and the transition rate. This may be done in the following way. Suppose that the target particle is located inside the volume V. The time spent by the projectile in the volume V is L/v where V is taken to be a cube of side L. The 'number of transitions' is then given by

$$N = \frac{RL}{v},$$

while the cross-section is given by

$$N = \frac{\sigma}{\text{area}} = \frac{\sigma}{L^2}$$

so that $\quad \sigma = \dfrac{L^2 RL}{v} = \dfrac{VR}{v}.$

In the c.m.s. $v = v_1 + v_2$, and if q is the initial momentum of each particle in the c.m.s., then we can write

$$\frac{1}{V} = \left[q \left\{ \frac{1}{e_I} + \frac{1}{e_T} \right\} \right]^{-1},$$

where e_I and e_T are the incident and target total c.m.s. energies. Substituting in the transition rate formula we then obtain

$$\sigma = \frac{g^2}{4\pi} \frac{p}{q} \frac{e_{F_1} e_{F_2}}{E^2},$$

where the e_F are the fermion energies, p are the final-state momenta, and we have written $E = e_1 + e_2 = e_I + e_T$ for the total c.m.s. energy. At high energies

$$\frac{p}{q} \frac{e_{F_1} e_{F_2}}{E^2} \to 1 \quad \text{so that} \quad \sigma \simeq \frac{g^2}{4\pi}.$$

If we use as a unit of cross-section $\sigma_0 = (m_\pi)^2 \simeq 2 \times 10^{-24} \text{ mm}^2$, then we have that at high energies

$$\sigma \sim \sigma_0 \frac{1}{4\pi} (gm_\pi)^2.$$

If we now insert into this formula the value of, for instance, the cross-section for the process $\pi^- p \to \Lambda^0 K^0$ at high energies, which is approximately one millibarn, we obtain

$$g \sim (m_\pi)^{-1}.$$

We note that, as expected, this coupling constant has dimensions L^{+1}. If we wish to compare with the coupling constant for the decay of a boson into two fermions, which is dimensionless, then we must divide by the appropriate unit of length. We may take this to be the pion Compton wavelength $\hbar/m_\pi c = (m_\pi)^{-1}$ in our units. Thus we arrive at a value for $g^2/4\pi \simeq 0 \cdot 1$. Such a coupling constant is characteristic of the strong interactions, and in extreme contrast to the value of about 10^{-15} which is characteristic of the weak decays.

Chapter 5
Discovery of the strange particles

5.1 V⁰ particles

As with the π-mesons all the early work on what have been called the 'strange' particles was done by means of cosmic-ray studies, using cloud chambers at sea level and at mountain altitudes, and using nuclear emulsions flown in high-altitude balloons. The first example of a particle other than those we have already discussed was reported by Leprince-Ringuet in 1944. A secondary cosmic-ray particle, which crossed the cloud chamber, produced a delta ray, or recoil electron, having substantial energy and emitted at a measurable angle. From the measured curvatures of the tracks in the magnetic field, and the scattering angle, it was possible to determine the mass of the incident particle, which was found to be 500 ± 50 MeV$/c^2$. It is now clear that this particle must have been a K-meson, but at the time, when even the pion had not been identified, the significance of this single event was not clear.

   The first clear examples of the new particles were observed in 1947 by Rochester and Butler at the University of Manchester. In these events the decays of the particles were observed, allowing a more convincing conclusion than any that could be obtained from observations on a single track, even when a delta ray was produced. Rochester and Butler operated a cloud chamber in a magnetic field and triggered it when an arrangement of Geiger counters near the chamber detected a penetrating cosmic-ray shower. In the course of about one year of operation at sea level they obtained two photographs out of fifty which showed examples of what were called V-particles. These photographs are shown in Figure 37. In the first of these the V appears to have originated just below a 30 mm lead plate across the centre of the chamber. A number of possible explanations were tested for these events.

(a) It is possible that the V might represent a very-large-angle scatter of one of the particles at the vertex. In such a case one might expect to see a short track due to the recoiling nucleus, and such a hypothesis could be quantitatively tested by measurements of the curvatures and drop density of the tracks. Such a test showed that this hypothesis was untenable. In addition, in later work, when an increasing number of such events was collected, it became clear that such possible 'scatters' did not take place any more frequently in the solid metal plates used in the cloud chambers than in the gas itself.

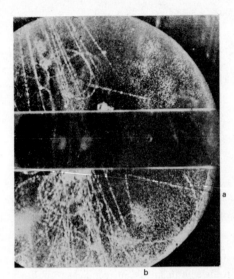

Figure 37 The first V⁰ particles observed by Rochester and Butler (1947) in a counter-controlled cloud chamber. In the left-hand photograph the V⁰ (a, b) originates just below the lead plate across the chamber. In the right-hand photograph the V⁰ (c, d) originates near the top right perimeter

(b) Another possibility was that the V of Figure 37 might be the decay of a charged particle at the apex point. However this too could be shown to be inconsistent with the measured quantities.

(c) The third possibility was that the event represented the decay of a neutral particle formed in the interaction of one of the cosmic-ray particles in the lead plate, the neutral particle decaying at the apex point into a pair of oppositely charged particles. Such a hypothesis was found to be entirely consistent with the data.

Similar cloud-chamber experiments were continued by the Manchester group at high altitudes and also by a number of other groups including that of R. W. Thompson in the USA. Thompson achieved a high degree of precision in momentum and angle measurement. With this increased precision, and with a growing number of events, the pattern of the data was gradually resolved. It is of interest to look more closely at the nature of the problem of identifying such particles. In the case where the origin of the neutral V-particle is known, it is possible to check whether the neutral particle and the two charged particles are *coplanar*. If this condition is found to hold, it is very unlikely that any neutral particles are present other than the primary. Secondly, it is possible to check the conservation of transverse momentum in a direction at right angles to that of the primary. Further analysis is more difficult unless it is possible to identify the nature of the charged particles taking part in the decay. This may be done by a combination of curvature and drop-density measurements in the cloud chamber, or multiple-scattering and grain-density measurements in a nuclear emulsion. Unfortunately, this was only possible in very few cases and in none of the very first events. However a technique was introduced by the Manchester group which enabled them to show that two kinds of V-particle decay existed, having different decay products and Q-values. For each event they calculated a quantity

$$\alpha = \frac{p_+^2 - p_-^2}{p^2} = \frac{m_+^2 - m_-^2}{M^2} + 2p^* \left(\frac{1}{M^2} + \frac{1}{p^2} \right)^{\frac{1}{2}} \cos \theta^*,$$

where p_+ and p_- and m_+ and m_- are the momenta and masses of the decay particles, p is the momentum and M the mass of the incident neutral particle, and θ^* is the centre-of-mass angle between the incident and decay particles, for which p^* is the centre-of-mass momentum. In the centre-of-mass system the average value of $\cos \theta^*$ is zero, so that we can write

$$\langle \alpha \rangle = \frac{m_+^2 - m_-^2}{M^2}.$$

Thus the distribution of α will show a peak for any given decay of a primary particle, of definite mass, into two given decay products. In fact the Manchester group were able to show that the distribution in α had at least two peaks. The results obtained by this group, and by Thompson and other workers, established unambiguously the existence of the two decays

$$\Lambda^0 \to p + \pi^-$$

and $K^0 \to \pi^+ + \pi^-$.

We may note that once the product particles had been identified, the mass of the neutral primary could be obtained from the relationship

$$M^2 = (E_+ + E_-)^2 - (\mathbf{p}_+ + \mathbf{p}_-)^2,$$

where E_+ and E_- are the total energies of the decay particles. This expression for the missing mass, which is a Lorentz-invariant quantity, will also be found useful when we discuss the strongly decaying resonances.

Following the early cloud-chamber work much success in this field was achieved in nuclear-emulsion studies. This was so for the following reasons:

(a) The emulsions could be flown at high altitudes, collecting data for prolonged periods. The flux of the new particles was much greater at such altitudes than at sea level or at mountain altitudes.

(b) The stopping power of the emulsion is high, but even so it is possible to measure very short tracks. For instance, the range of a proton of 14 MeV is 1 mm, but measurements may be made to within a few microns.

(c) It often proved possible by flying stacks of plates or blocks of emulsion to observe *both* the origin and the decay of the new particles.

As with the study of the π-mesons, the most precise determinations of the masses, lifetimes and decay modes of these particles, as well as the study of their production and the determination of their other quantum numbers, had to await their production by accelerators. In the more detailed discussion of the new particles in the rest of this chapter we shall use the most significant data available, regardless of date.

5.2   Charged K-mesons

In the previous section we have seen evidence for the existence of what is now called a K^0 meson, decaying into a pair of oppositely charged pions. It was shortly after this discovery that evidence was also obtained for the existence of charged K-mesons. The identification of the charged kaons is easier for the positive than for the negative particle; these particles turn out to have strong interactions with nucleons so that the negative kaon is attracted by nuclei, and interacts before it can decay, while the positive kaon, due to the Coulomb repulsion, is more frequently able to decay beyond the range of the strong forces. We shall therefore first consider the decay modes of the positive kaons.

(a) The most striking decay mode is that into three charged pions

$$K^+ \to \pi^+ \pi^+ \pi^-.$$

The particle decaying in this way was originally referred to as the τ-meson. The first examples of this mode were observed in nuclear emulsions exposed to the cosmic radiation. In the early days of emulsion work, the emulsions were not sensitive to tracks of minimum ionization. In 1948 so-called electron-sensitive emulsions became available with a resulting series of new discoveries concerning elementary particles. A particularly striking example of this decay is shown in Figure 38. In this case the distinctive three-prong star results from a track arising from a disintegration in the emulsion due to a cosmic-ray particle of high energy. The unusual feature of this example is that two of the three secondary prongs from the decay can be identified by their decays, or interactions, in the emulsion. One of them comes to rest in the emulsion and decays, and the secondary decay particle is itself observed to decay into an electron. In fact this particle, and the one which leaves the emulsion without decaying or interacting, may be identified as π-mesons by means of the grain density, multiple scattering and range. The third particle can be seen to come to rest and interact in the emulsion, producing a track identifiable as a proton, suggesting that it is in fact a π^- meson. The three secondary particles can be shown to be coplanar, indicating the absence of any neutrals in the decay. For such a decay it is possible to obtain fairly precise values for the kaon mass, which could be shown to be about 494 MeV/c^2.

(b) Corresponding to the decay mode of the neutral meson into two pions, we have a two-pion decay mode of the charged K

$$K^+ \to \pi^+ \pi^0.$$

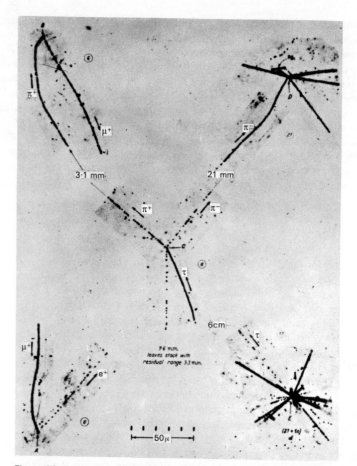

Figure 38 Production of a τ-meson in a nuclear emulsion exposed to the cosmic radiation. The τ emerges from a 'star', slows down to rest and decays to three tracks. One of these is identified as a π^+ by its $\pi \to \mu \to$ e decay. Another is identified as a π^- by its absorption by a nucleus in the emulsion to give another 'star' (Powell, Fowler and Perkins, 1959, p. 63)

The particle decaying in this way was originally known as the θ-meson. The primary in this decay was first identified in emulsions as having the K-mass by observations on its track. The charged secondary could be identified as a π-meson, and was found to have a unique momentum in the centre-of-mass system of the decaying particle. As in the case of the $\pi \to \mu$ decay this indicates that only one neutral particle can be present, although it was not possible in the early work to establish whether the neutral was a π^0 or a gamma ray. Increasing data provided a number of events in which a direct pair from π^0 decay (i.e. $\pi^0 \rightarrow e^+ e^- \gamma$) was observed, thus removing the ambiguity.

For this decay, and for all other decays involving π^0 mesons or gamma rays, a substantial advance in technique was afforded by the development of bubble chambers containing liquids of short radiation length. In such chambers the probability of converting one or even both gamma rays from π^0 decay is high. Particularly useful in this respect have been the studies carried out using bubble chambers containing liquid xenon, which has a radiation length of only 30 mm. For the decay under discussion, if the K-meson is brought to rest in a liquid-xenon chamber then in many cases the event will show two electron pairs from the π^0 decay, enabling a full reconstruction to be achieved.

(c) The last of the two-particle decays to be established was the process

$$K^+ \rightarrow \mu^+ \nu.$$

In fact this mode turns out to be the most common decay channel for the K^+. The reason for the difficulty in establishing the nature of this decay was that the charged secondary has a rather high momentum, making it more difficult to identify in a nuclear emulsion unless long tracks are available. The first events of this nature were identified, by means of a multi-plate cloud chamber, by the Ecole Polytechnique group in Paris. The cloud chamber used was in fact a double chamber in which the upper chamber was without plates but placed in a magnetic field. In this chamber the momentum of an incident particle could be measured. The lower chamber contained a series of metal plates, but no magnetic field was present. The multi-plate chamber gave the residual range for particles which stopped within it, and the combination of these data frequently gave the mass of the particle with a precision adequate to identify it as a K-meson.

In the cases in which the secondary particle emitted by the stopped K was in such a direction as to pass through a number of plates, it was sometimes possible to determine its nature by means of observations on its residual range and rate of energy loss in passing through the plates. In addition, the fact that such charged secondaries were of uniformly high momentum suggested a two-body decay and detailed kinematic analysis indicated that the associated neutral particle should have very small or zero mass. The absence of associated showers in the plates established that the neutral particle could only be a neutrino.

(d) Corresponding to the charged three-pion decay we might expect, on the basis of charge independence, that we should also have the mode

$$K^+ \rightarrow \pi^+ \pi^0 \pi^0.$$

For the $K^+ \rightarrow \pi^+ \pi^0$ decay, the π^+ has a unique kinetic energy of about 108 MeV. The first example of the above three-pion decay was identified by the presence of a slow π^+ having an energy of only about 14 MeV. In fact for this decay mode we would expect a continuous spectrum of π^+ mesons up to a maximum energy of about 53 MeV. As with the other processes involving π^0 mesons, the clearest evidence can be obtained from the xenon bubble chamber where pairs from all four gamma rays may sometimes be observed.

(e) The mode

$$K^+ \rightarrow \mu^+ \pi^0 \nu$$

was first identified in a similar way to that above in cases where the primary tracks could be identified as having the kaon mass, while the secondary muons were observed to have a variety of energies, indicating that at least two neutrals were present. As usual, in about 1 per cent of cases the π^0 decays into a gamma ray plus two electrons, so that a pair is observed starting from the vertex along with the muon. Such events have been found to be inconsistent with the decay scheme

$$K^+ \rightarrow \mu^+ e^+ e^- n_1,$$

where n_1 has a unique mass, but do fit the hypothesis

$$K^+ \rightarrow \mu^+ \pi^0 n_2,$$

where the mass of n_2 is less than 75 MeV/c^2, and consistent with zero, so that n_2 must be either a gamma ray or a neutrino. If we know that the K-particle is a boson, then n_2 must be a neutrino. This of course assumes that the kaon observed to decay in this way is the same particle as the kaon which decays into only π-mesons. Again a fuller identification of the decay has been made in the xenon bubble chamber.

(f) The mode

$$K^+ \rightarrow e^+ \pi^0 \nu$$

has been established in several experiments in which it has proved possible to identify the fast positively charged particle as a positron. The energy of the positron is not unique thus requiring the presence of two neutral particles, one of which has been observed to be π^0 by means of detection of its gamma rays.

It is also possible that other very rare decay modes may exist such as

$$K^+ \rightarrow \begin{cases} \pi^+ \pi^- e^+ \nu \\ \pi^+ \pi^+ e^- \nu \\ \pi^+ \pi^- \mu^+ \nu \\ e^+ \nu \\ \mu^+ \nu \gamma \end{cases} \qquad K^+ \rightarrow \begin{cases} \pi^+ \pi^0 \gamma \\ \pi^+ \pi^+ \pi^- \gamma \\ \pi^+ e^+ e^- \\ \pi^+ \mu^+ \mu^-. \end{cases}$$

The relative frequencies of the different decay modes, usually referred to as the branching ratios, have been determined in a large number of different experiments and are summarized for the principal modes in Table 2.

Table 2

Decay mode	Frequency/per cent
$K^+ \rightarrow \mu^+\nu$	$63\cdot4 \ \pm0\cdot5$
$\pi^+\pi^0$	$21\cdot0 \ \pm0\cdot3$
$\pi^+\pi^+\pi^-$	$5\cdot6 \ \pm0\cdot1$
$\pi^+\pi^0\pi^0$	$1\cdot71\pm0\cdot08$
$\mu^+\pi^0\nu$	$3\cdot41\pm0\cdot22$
$e^+\pi^0\nu$	$4\cdot79\pm0\cdot18$

5.3 Mass and lifetime for charged K-mesons

5.3.1 *Mass*

The most precise determination of the K-meson mass has been made by means of a reaction involving sigma particles, which we have not yet discussed. This determination will be mentioned in a later section. An alternative method, which yielded an accurate value, consisted of a measurement of the momentum of the K-particles by magnetic analysis combined with a measurement of their energy from their range in a nuclear-emulsion stack. The arrangement is shown in Figure 39. A target placed in the internal circulating beam of the Bevatron yielded K-mesons, and other particles, which were allowed to pass through a set of magnetic quadrupole lenses (see Chapter 1) and a bending magnet before entering an emulsion stack. The position at which they entered the stack was thus a function of the particle momentum. The momentum–position relationship was established by measuring

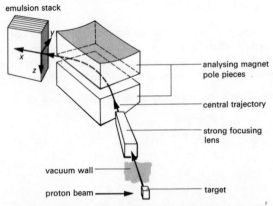

Figure 39 Experimental arrangement for the determination of the kaon mass, using magnetic analysis to measure the momentum and energy (Peterson, 1957)

the residual range of protons entering the stack over the region of interest. These measurements yielded a value for m_K of (493 ± 1) MeV/c^2.

5.3.2 Lifetime

A most precise value of the lifetime for the charged K-meson has been obtained by time-of-flight measurements at Brookhaven. An electrostatically (see Chapter 1) separated K-meson beam was allowed to pass through scintillation counters, the distance between which could be set at 620, 1290 and 1950 in (15·75, 32·77 and 49·53 m). In addition to the electrostatic separation, the K-mesons were identified by means of two differential Cerenkov counters. Measurements were made of the attenuation of the K-meson flux at momenta of 1·6 and 2·0 GeV/c. A precise check of the geometric efficiency of the beam was obtained by using stable particles, protons or antiprotons, while the effect of the small amount of material in the beam was corrected for by studying the transmission as a function of the thickness of material, and extrapolating to zero absorber. A prime objective of these measurements was a precise comparison of the lifetime for the positive and negative particles, which could be obtained by changing the polarity of the bending magnet. The number of particles remaining after travelling a distance L is then given in terms of the mean lifetime τ by the expression

$$N = N_0 e^{-Lm_0/p\tau},$$

where N_0 is the original number of particles, m_0 the particle rest mass in MeV/c^2, p the particle momentum in MeV/c and L/τ is in units of the velocity of light. This relationship takes into account the dilatation of the particle lifetime due to its motion. The data obtained are consistent with a pure exponential decay to better than 1 per cent at the largest distance. The absolute lifetime of the K$^+$ is found to be $12·265 \pm 0·035$ ns (1 ns = 10^{-9} s). The ratio of the lifetimes τ_+ and τ_- for the positive and negative particles is found to be

$$\left(\frac{\tau_+}{\tau_-} - 1\right) = -0·0010 \pm 0·0017,$$

completely consistent with equality. We shall refer to this result in a later section when we deal with invariance under the so-called *CPT* transformation.

5.4 Neutral K-mesons

(a) $K^0 \rightarrow \pi^+ \pi^-$.

We have already seen that the original observations of V-particles indicated the existence of a particle decaying into two charged π-mesons. Whenever the origin of such an event was observed it was found that the neutral line of flight was co-planar with the secondaries, indicating that no other neutral particles were present. The early data established that the mass of the primary was 492 ± 3 MeV/c^2 yielding a Q-value for the decay of 214 ± 3 MeV. The lifetime for these particles decaying in this way was best established in the early data using kaons produced and decaying in a 12 in (305 mm) propane bubble chamber. The momentum of the K^0s was established from the direction and curvatures of the decay pions and used to determine the time of flight in each case. The distribution in time of flight was found to be closely exponential, at least for short times, yielding a value for the lifetime of $(0.95 \pm 0.08) \times 10^{-10}$ s.

(b) $K^0 \to \pi^0\pi^0$.

This decay mode is to be expected, on the basis of charge independence, as an alternative to that involving two charged pions. As in previous situations where π^0 mesons are present it is most clearly observed in the liquid-xenon bubble chamber, where in favourable circumstances electron pairs from all four decay gamma rays may be observed. The lifetime for this decay mode was measured in early experiments by an interesting method in which the decay of K^0s produced from the target of an accelerator was measured as a function of the distance they had travelled before the decay took place. The apparatus is shown schematically in Figure 40. The arrangement of collimators is such that only gamma rays emitted

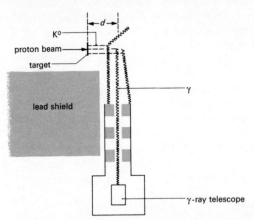

Figure 40   The arrangement used by various workers to study the lifetime of K^0 mesons produced by a proton beam in an accelerator. The γ-intensity was studied as a function of d (Franzinetti and Morpurgo, 1957)

near 90° from the proton beam direction are detected. Gamma rays arising directly from the target, or from mesons produced in the target, are prevented from reaching the gamma-ray detector by a thick lead screen. The gamma-ray detector consisted of an arrangement of scintillation counters, a lead convertor and a water Cerenkov counter. π^0 mesons arising from the decay of charged K-mesons give negligible contribution compared with those from neutral kaons due to the fact that the lifetime for the charged K is about 100 times greater than that for the short-lived neutral kaon. The counting rate in the gamma-ray telescope, for a given incident proton beam on the target, was then studied as a function of the distance

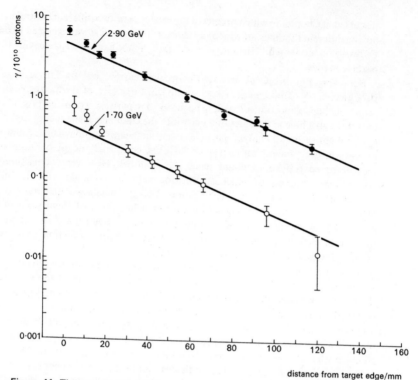

distance from target edge/mm

Figure 41 The counting rate in the γ-ray telescope of Figure 40 as a function of the distance d from the target edge. The slope of the exponential decay curves yields the K^0 lifetime

of the decay point from the target edge. At distances greater than about 15 mm the counting rate was found to fall exponentially with distance (see Figure 41). In order to determine the lifetime from a measurement of the gradient of this curve it is necessary to know the momentum distribution of the K^0 particle. This can only be obtained from studies of proton–nucleus collisions, of a similar energy, in an experiment in which the K^0 particles are observed to decay in the charged mode, and by assuming that the energy spectrum of such neutral kaons is the same for both charged and neutral decays.

The ratio of these decay modes was found to be

$$K^0 \rightarrow \pi^+ \pi^- : K^0 \rightarrow \pi^0 \pi^0 = 69 : 31.$$

Even in the early observations a small percentage of so-called anomalous K^0 decays which could not be interpreted as decays into two pions were observed. We may anticipate the result of the Gell-Mann–Nishijima theory (which will be discussed in Chapter 6) which predicted the existence of two different neutral kaons, having different lifetimes, of which the shorter-lived particle was expected to decay by two-pion modes while the longer-lived particle was expected to decay by three-particle modes.

Following this proposal an experiment was carried out by Lande *et al.* at Brookhaven. A cloud chamber was set up, at a distance of six metres from the target of the Cosmotron, in such a position that particles from the target could pass through a hole in the shielding mass into the chamber. A magnet placed in the line of flight swept all charged particles from the beam. At this distance and with primary protons having an energy of 3 GeV a completely negligible number of short-lived neutral kaons could reach the chamber. However twenty-three V^0 events were observed, of which all but one were not coplanar with the line of flight from the target to the decay point. Also it was shown that none of the secondaries from these decays was a proton, while most of them were in fact lighter than K-mesons, that is were pions, muons or electrons. This experiment thus established the existence of a long-lived neutral kaon, of lifetime of the order of 10^{-8} s, decaying into at least three particles.

An alternative piece of evidence pointing to the existence of long-lived neutral kaons was provided by the application of another prediction of the Gell-Mann theory. This was that the K-particles must always be produced *along with* another so-called 'strange' particle such as the Λ-particle, decaying into a proton and a π-meson, which we have mentioned at the beginning of this chapter. In an experiment in which such Λ-particles were produced by π mesons in a propane bubble chamber, the K^0 was only observed to decay in the chamber in about 50 per cent (after correction) of all cases. This indicates that in the other 50 per cent of cases the K^0 lived for a time sufficiently long to escape from the chamber. By methods similar to those already described for the short-lived K^0 particles, the following decay modes have been identified:

$$K^0 \rightarrow \begin{cases} \pi^\pm e^\mp \nu \\ \pi^\pm \mu^\mp \nu \\ \pi^+ \pi^- \pi^0 \\ \pi^0 \pi^0 \pi^0. \end{cases}$$

In addition, it has been found that a very small proportion of the long-lived neutral kaons decay into *two* pions. This feature will be discussed when the K^0 particles and their curious properties are treated more fully in Chapter 8.

5.5 The Λ^0 hyperon

In section 5.1 we saw that the original observations of V-particles indicated the existence of a heavy particle, decaying into a proton and a π^- meson, in addition to the neutral K-meson which decays into two pions. This particle is known as the Λ-*hyperon*.

We may summarize the terminology as follows:

Nucleons are the proton and neutron, and antinucleons the antiproton and anti-neutron.

Hyperons were originally defined as particles heavier than the nucleons. However, as we shall see, this definition is no longer adequate and we must *add* the requirement that hyperons have baryon number equal to 1. The antihyperons are particles of mass greater than the nucleon with baryon number equal to −1. We shall see that, like the nucleons, hyperons are fermions and have half-integral spin.

Mesons are defined as strongly interacting particles with baryon number 0, and we shall see that they are all bosons having integral spin.

All the above interact via the strong interactions and are known collectively as *hadrons*.

Leptons are particles subject only to the weak and electromagnetic interactions. The leptons turn out to be all fermions lighter than the π-meson, as suggested by the name. In fact the only such particles known are the muon, the electron and the neutrino and their antiparticles.

The only particle which does not fall into this classification is the photon, which is a boson subject only to the electromagnetic interaction.

Returning to the Λ^0 hyperon, the decay mode first observed was

$$\Lambda^0 \to p\pi^-,$$

established by observations in cloud chambers and nuclear emulsions, which identified the decay tracks as a proton and a π^- meson. The two-body nature of the decay was determined by observations on the coplanarity of the neutral particle and the two charged tracks, in cases where the origin of the V^0 could be observed. It should be noted that, in practice, the coplanarity measurements need to be very precise in order to exclude the existence of a third particle. In addition, the mass of the neutral could be established by determining the quantity

$$(E_p + E_{\pi^-})^2 - (\mathbf{p}_p + \mathbf{p}_{\pi^-})^2,$$

where E and $\mathbf{p}$ are the total energy and momentum of the particle. The distribution of this quantity, for a typical set of V^0 particles produced in a bubble-chamber experiment, is shown in Figure 42. The sharp peak gives a measurement of the mass, and provides additional evidence that there are only two decay particles. The mass is found to be $1115 \cdot 58 \pm 0 \cdot 10$ MeV/c^2. The lifetime has been measured by plotting the distribution of the delay between production and decay, in the rest system of the Λ^0 particle, for events observed in a bubble chamber, where the production and decay of the particle were seen and the measurements allowed a good determination of the Λ^0 momentum, necessary to make the transformation from the laboratory to the particle centre-of-mass. Such measurements yield a distribution well fitted by an exponential decay corresponding to a mean life of $(2 \cdot 51 \pm 0 \cdot 04) \times 10^{-10}$ s.

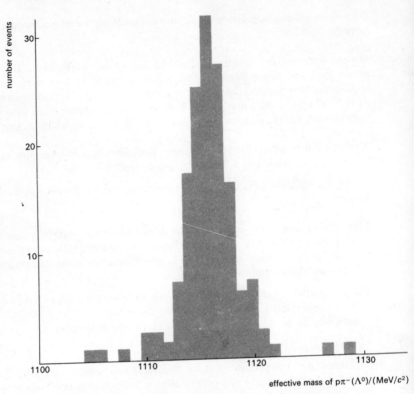

Figure 42 Distribution of the effective mass for a set of V⁰ particles having effective mass near the Λ⁰ mass. The data are from a sample of V⁰ formed in K⁻p interactions at 6 GeV/c

On the basis of charge independence we should also expect the decay

$$\Lambda^0 \to n\pi^0.$$

This process can be observed only by detecting the pairs from the π^0 decay gamma rays. In occasional cases where the Λ^0 decays in the hydrogen bubble chamber, the neutron may collide with a proton in the liquid to produce a recoil track. From observations of events in which a Λ^0 is expected to be present because of the existence of a K^0 (see Chapter 6 concerning 'associated production'), such electron pairs, and sometimes proton recoils, have been observed. It might be possible that the Λ^0 should decay via the mode

$$\Lambda \to n\gamma.$$

This is however excluded on a statistical basis by plotting the energy distribution of the hypothetical gamma rays for a number of events. For such a mode the distribution should be monoenergetic.

For the decay

$$\Lambda \to n\pi^0,$$

we expect a flat distribution of gamma-ray energy between 32 and 134 MeV (compare the process $\pi^- p \to n\pi^0$ for absorption at reat; see section 2.5). Such a flat spectrum is in fact observed.

The ratio of the decay modes $(\Lambda \to p\pi^-)/(\Lambda \to n\pi^0)$ is found to be almost exactly 2. We shall see later that this ratio is an example of the rule that the isotopic spin changes by $\frac{1}{2}$ in weak-decay processes.

5.6 The $\Sigma^\pm$ hyperons

The first evidence for these particles came from tracks observed in nuclear emulsions exposed to cosmic rays (Figure 43). Observations on the grain density and multiple scattering of the tracks indicated masses of about 1200 MeV/c^2. The decay is into a single charged particle; with accumulation of data it became clear from the unique energy of the decay particle, in the decay centre-of-mass, that only one additional neutral particle could be present. By the usual methods three principal modes were identified;

$$\Sigma^+ \to p\pi^0,$$
$$\Sigma^+ \to n\pi^+,$$
$$\Sigma^- \to n\pi^-.$$

In addition, decay modes involving leptons are now known to occur with very low frequency. Examples of such modes are

$$\Sigma^+ \to \begin{cases} p\gamma \\ n\pi^+\gamma \\ \Lambda e^+ v \end{cases} \qquad \Sigma^- \to \begin{cases} ne^- v \\ n\mu^- v \\ \Lambda e^- v \\ n\pi^- \gamma. \end{cases}$$

The branching ratio $(\Sigma^+ \to p\pi^0)/(\Sigma^+ \to n\pi^+)$ is found to be nearly equal to 1.

The most precise values for the masses of the charged Σ-hyperons, and also of the K^- meson, have been obtained from range measurements in nuclear emulsions on the charged products of the decays $\Sigma^+ \to p\pi^0$ and $\Sigma^+ \to n\pi^+$ and the reactions

$$K^- p \to \Sigma^+ \pi^-, \qquad\qquad \textbf{5.1}$$
$$K^- p \to \Sigma^- \pi^+. \qquad\qquad \textbf{5.2}$$

The Σ^+ mass was obtained from the proton range in the decay $\Sigma^+ \to p\pi^0$. The K^- mass could then be obtained from the Σ^+ range in **5.1**. The mass difference between the Σ-particles of opposite charge could be obtained from the range difference in these particles in **5.1** and **5.2**. For events in which the K^- is captured at rest, and

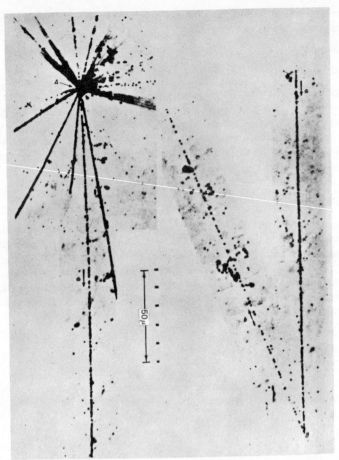

Figure 43 An example of the production and decay of a Σ-particle in a nuclear emulsion. The Σ emerges from a large, cosmic-ray induced star. Its track is continued on the right of the picture, and decays at the lower right (Powell, Fowler and Perkins, 1969, p. 359)

in which the Σ^+ decays at rest, these measurements should yield the appropriate masses with high precision. However detailed study of the process **5.2** indicated a definite failure to balance momentum on the basis of the masses calculated as just described. The reason for this apparent lack of momentum conservation is explained by a difference between the rate of energy loss of slow negative and positive Σ-particles of the same velocity (Barkas *et al.*). A reassessment of the range–energy relationship for positive and negative particles yields a best fit which removes the momentum conservation anomaly and gives masses as follows:

$M_{\Sigma^+} = 1189{\cdot}35 \pm 0{\cdot}15$ MeV/c^2 (range measurement of proton in $\Sigma^+ \to p\pi^0$),

$M_{K^-} = 493{\cdot}7 \pm 0{\cdot}3$ MeV/c^2 (range of Σ^+ in reaction **5.1**),

$M_{\Sigma^-} = 1197{\cdot}6 \pm 0{\cdot}5$ MeV/c^2 (ranges of the pions in reactions **5.1** and **5.2**).

The mean lives of the Σs have been determined from observations on the distribution of their track lengths before decay. It is interesting to note that the angle through which a particle is bent by a magnetic field is independent of both particle momentum and of its dip with respect to the axis of observation, which is taken to be along the direction of the field (see Figure 44). Thus, for a particle of constant

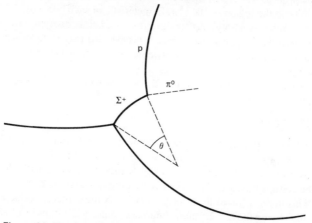

Figure 44 Illustration of the decay of a Σ^+ particle to proton and neutral pion showing the angle θ discussed in the text

mass, a plot of the distribution of θ should show an exponential distribution, eliminating the need to transform to the centre-of-mass system of the decaying particle. The lifetimes are found to be

Σ^+ : $(0{\cdot}810 \pm 0{\cdot}013) \times 10^{-10}$ s,

Σ^- : $(1{\cdot}65 \pm 0{\cdot}03) \times 10^{-10}$ s.

As might be expected, on the basis of charge independence and the existence of two channels for the decay of the positive Σ-hyperon and only one for the decay of the negative Σ-hyperon, the lifetime τ_{Σ^-} is found to be twice τ_{Σ^+}. As with the Λ, the lifetimes are short so that the distance between production and decay is normally short, except for particles of very high momentum which benefit from time dilatation. This feature made it difficult to establish accurate values for the mass from observations on the tracks in the early work.

5.7 The Σ^0 hyperon

It was a prediction of the Gell-Mann–Nishijima scheme (see Chapter 6) that the Σ-hyperons should have isotopic spin equal to 1, implying an i-spin triplet with I_3 changing by one unit between members and thus the existence of a neutral Σ-particle.

A consequence of this scheme was also that the decay

$$\Sigma^0 \to \Lambda^0 \gamma,$$

accessible to the neutral Σ, would not be inhibited by strangeness conservation as are the charged Σ-decays described in the previous section. It is thus possible for this decay to proceed via electromagnetic interactions, so that we might expect the lifetime to be in the region of 10^{-15} s characteristic of such processes. For such a lifetime the Σ^0 will travel only a rather short distance before decaying; this makes its identification difficult, particularly since it decays into two particles which are themselves uncharged. Such a decay thus corresponds to a neutral vertex in a bubble chamber.

The first convincing evidence for the Σ^0 was provided by an event of the following kind

$$\pi^- p \to \Lambda^0 K^0 + (\text{neutral}),$$

where the measurements of the missing momentum and energy identified the unobserved object as a gamma ray. Subsequently, events have been observed in which an electron pair from the gamma ray is produced along with the other particles. The event is then kinematically overdetermined, and the Σ^0 mass may be calculated. This mass is found to be 1193 MeV/c^2. A more precise value may be obtained by measuring the $\Sigma^- - \Sigma^0$ mass difference. When a Σ^- stops in a hydrogen bubble chamber it may interact with a proton according to the processes

$$\Sigma^- p \to \Lambda^0 n,$$
$$\Sigma^- p \to \Sigma^0 n \to \Lambda^0 \gamma n,$$

if the Σ^- is heavier than the Σ^0. The first of these processes will yield mono-chromatic Λ-particles of 36·9 MeV. The second process will produce a continuous spectrum of Λ-particles. The $\Sigma^- - \Sigma^0$ mass difference can then be obtained from the upper and lower limits for the Λ-energy. The difference $M_{\Sigma^-} - M_{\Sigma^0}$ is found to be 4·9 MeV/c^2.

The lifetime of the Σ^0 has only been determined to be $\leqslant 10^{-14}$ s, the best data being from nuclear emulsions. Theoretical calculations indicate a lifetime of about 5×10^{-17} s.

5.8 The Ξ^- hyperon

This particle is also known as the cascade hyperon, since it decays into a pion

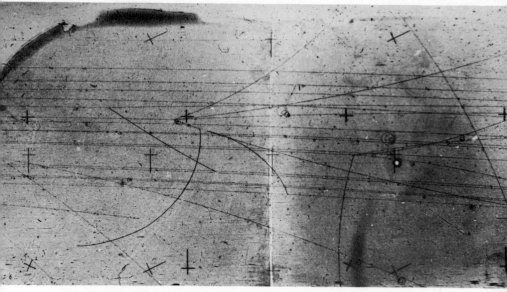

Figure 45 An example of Ξ^- production and decay in the British 1·5 m bubble chamber exposed to a 6 GeV/c K$^-$ beam from the CERN proton synchroton.
The Ξ^- emerges as the lowest of four tracks from the interaction in the left-hand half of the picture and rapidly decays to a Λ^0 and a π^-

and a Λ-hyperon. In a cloud or bubble chamber the topology is that shown in Figure 45, where the Λ originates from the decay point of the negative particle and the charged member of the decay can be identified as a π^- meson,

$$\Xi^- \to \Lambda^0 \pi^-.$$

The first example of such a process was reported by the Manchester group, from a cloud-chamber photograph in which the V^0 line of flight passed through the negative decay point. In this case, however, it was not possible to identify the V^0 as a Λ-particle. A very large number of such cascade particles have been observed in bubble-chamber experiments, using incident K$^-$ beams from accelerators, and the properties of the Ξ^- have been studied in detail, particularly using chambers containing heavy liquids. Since both decay products may be observed, the mass may be determined with quite high precision and the best value currently available is 1321·2±0·2 MeV/c^2. We note that this mass is such that decay to $\Sigma+\pi$ is energetically just forbidden.

The lifetime obtained by methods similar to that for the Σ^- is found to be $(1·74±0·05)\times10^{-10}$ s.

5.9   The Ξ^0 hyperon

Another prediction of the Gell-Mann–Nishijima scheme was that the cascade hyperon should be an i-spin doublet, thus suggesting the existence of a neutral cascade particle (compare the proton–neutron i-spin doublet). We might expect such a cascade particle to decay via the process

$$\Xi^0 \to \Lambda\pi^0,$$

illustrated in Figure 46, where we have assumed production via the reaction

$$K^-p \to \Xi^0 K^0.$$

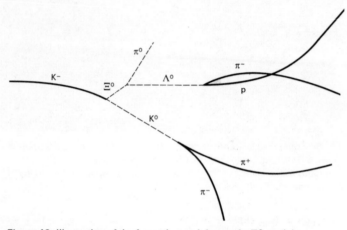

Figure 46   Illustration of the formation and decay of a Ξ^0 particle

As with the Σ^0 the decay is an all-neutral vertex, rendering more difficult the identification and study of this particle. However, unlike the Σ^0, we might expect that the Ξ^0 should have a lifetime comparable with that of the Ξ^-, so that it may travel an observable distance before decay. Many examples of processes which can only be fitted by the production and decay of a neutral cascade particle have now been observed. The best data can be obtained in cases where one, or even both, of the gamma rays from the π^0 in the decay produce electron pairs, as is the case quite frequently for the π^0s decaying in a heavy-liquid bubble chamber. Even where this is not so it may be possible to reconstruct the event. For instance in the process illustrated in Figure 46 for incident K-mesons of known energy we may proceed as follows:

(a) Attempt to identify the lower V^0 from measurements on its decay tracks. This may enable it to be identified as a K^0 meson of known momentum.

(b) We may now test the hypothesis that only one other neutral particle is produced in the reaction, by calculating the magnitude and direction of the momentum necessary for conservation at the primary vertex. If we also apply energy conservation at this vertex we may calculate the mass of the neutral particles in the usual way.

(c) We also attempt to identify the particle responsible for the upper V^0 in the figure, and in our example this may be identified as a Λ-hyperon whose momentum may be calculated in magnitude and direction.

(d) We may now test whether the calculated momentum vectors for the Λ and the Ξ^0 intersect in space. If this is found to be true, then for a two-particle decay of the Ξ^0 there will be a unique relationship between the emission angle and the momentum of the Λ. This relationship will provide a test of our hypothesis.

(e) In addition, if pairs from the π^0 decay gamma rays are observed yet a further check may be applied.

In practice, for such an event observed in a bubble chamber, the hypotheses for each charged and neutral vertex are tested individually and then an overall fit to the complete event is made.

The mass of the cascade zero is found to be $(1314 \cdot 7 \pm 1 \cdot 0)$ MeV/c^2 and its lifetime $(3 \cdot 0 \pm 0 \cdot 5) \times 10^{-10}$ s.

5.10 The Ω^- hyperon

This particle was predicted by Gell-Mann on the basis of SU(3) symmetry and

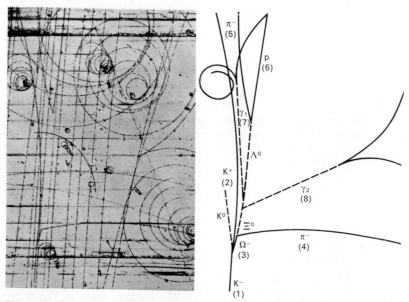

Figure 47 The first Ω^- particle to be observed (Brookhaven National Laboratory, 1964)

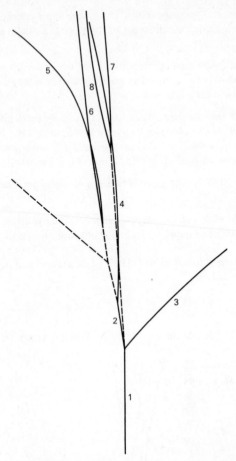

Figure 48 Outline of an Ω^- event produced by a 6 GeV/c K$^-$ meson in hydrogen. The original photograph is difficult to reproduce and confused by additional beam tracks. 1 is the incident K$^-$, 2 is the Ω^-, 3 is a K$^+$, 4 a π^-, 7 and 8 π^- and π^+ mesons from a K^0 decay, 5 and 6 π^- and p from Λ-decay (see text for details)

its discovery provided one of the most notable triumphs of this theory (see Chapter 10). The predicted mass was 1673 MeV/c^2 and the expected decay modes were

$$\Omega^- \rightarrow \begin{cases} \Xi^-\pi^0 \\ \Xi^0\pi^- \\ \Lambda K^-. \end{cases}$$

The first example of the production of this particle was found in a bubble-chamber photograph at Brookhaven in 1964. The event is shown in Figure 47 and is interpreted as

$$K^- p \to \Omega^- K^+ K^0.$$

The incident beam consists of K^- mesons having a momentum of 5 GeV/c. The K^0 from the production vertex is not observed to decay in the chamber and the event could be fully analysed only because both gamma rays from the π^0, produced in the decay chain of the Ω^-, produced electron pairs in the bubble chamber. The Ω^- decay was reconstructed according to the following scheme

$$\Omega^- \to \Xi^0 \pi^-$$

$$\quad \hookrightarrow \Lambda \pi^0$$

$$\qquad \hookrightarrow \gamma\gamma$$

$$\qquad \hookrightarrow p\pi^- \quad \hookrightarrow e^+ e^-$$

$$\qquad\qquad \hookrightarrow e^+ e^-.$$

Another rather complete example of Ω^- production and decay is illustrated in Figure 48 from an experiment to study the interactions of K^- mesons with protons at a momentum of 6 GeV/c carried out using film obtained with a beam from the CERN proton synchrotron. The reaction is

$$K^- p \to K^0 K^+ \Omega^-$$

$$\quad \hookrightarrow \pi^+ \pi^- \quad \hookrightarrow \Xi^0 \pi^-$$

$$\qquad\qquad \hookrightarrow \Lambda \pi^0$$

$$\qquad\qquad\quad \hookrightarrow p\pi^-.$$

Referring to the figure, it is possible to identify track 8 as a π^+ from momentum and bubble-density measurements. Tracks 7 and 8 can be reconstructed as the decay of a K^0, mass 502 ± 4 MeV/c^2, produced at the initial vertex of the event with momentum 3398 ± 55 MeV/c. Similarly, identifying track 6 as a proton, tracks 5 and 6 are fitted as the decay of a Λ, mass 1115 ± 1 MeV/c^2 and momentum 1355 ± 20 MeV/c. The line of flight of the Λ misses the production vertex by $4\cdot7 \pm 0\cdot7$ mm, and the negatively charged decay vertex by $5\cdot6 \pm 0\cdot8$ mm. Track 3 can be interpreted as a K^+, from bubble-density and momentum measurements, a proton or a π^+ being ruled out. The transverse momentum of track 4 is 265 MeV/c with errors of ± 2 MeV/c and ± 6 MeV/c, using the fitted and unfitted quantities respectively, and is thus greater than the maximum possible for any of the common decay modes of π^-, K^-, Σ^- or Ξ^-; but it is consistent with the decay $\Omega^- \to \Xi^0 \pi^-$, at the angle we observe, yielding a Ξ^0 momentum of 1635 ± 6 MeV/c. The lines of flight of the Ξ^0 and Λ are then found to intersect to within $0\cdot3 \pm 0\cdot5$ mm in space and give a satisfactory fit for

$$\Xi^0 \to \Lambda^0 \pi^0$$

at the neutral vertex. A multi-vertex fit to the four vertices, production, negatively charged decay, Λ-decay and neutral Ξ^0 vertex, yields a good fit for an Ω^- mass of 1666 ± 8 MeV/c^2.

A number of Ω^- events have now been observed, some showing decay into ΛK^-. The best value of the mass is $1672\cdot4 \pm 0\cdot6$ MeV/c^2, and of the lifetime, obtained from the track lengths and momenta of the observed particles, $(1\cdot3 \pm 0\cdot4) \times 10^{-10}$ s.

Chapter 6
The Gell-Mann–Nishijima scheme and conservation of strangeness

6.1 Strange-particle production and decay

We have seen in the previous chapter that the lifetimes for the decay of the K-mesons, the Λ-particle, the charged Σ-particles, the Ξ-particles and the Ω^- are all of the order of 10^{-10} s, while the long-lived K^0 has a lifetime of about 10^{-8} s. Following the discussion of Chapter 4 it is clear that these lifetimes correspond to decay by virtue of the weak interactions.

On the other hand, it became clear when the production of strange particles was studied, using accelerator beams, that the production cross-sections were quite large. For instance, it was found that at energies somewhat above threshold the cross-section for Λ production in $\pi^- + p$ reactions was ~ 10 mb. The experiments which yielded this data were performed using cloud chambers and nuclear emulsions, in the early stages, and later predominantly by bubble chambers. The first accelerator-produced strange particles were studied with the aid of the Brookhaven 3 GeV Cosmotron and later using the Berkeley 6 GeV Bevatron.

Again, referring to the discussions of Chapter 4, we see that cross-sections of this magnitude are characteristic of strong interactions. Although there are uncertainties in the calculation of the coupling constants, from either decay lifetimes or production cross-sections, the difference in magnitude between the strong- and weak-interaction coupling constants is so great that it would be extremely difficult to account for the lifetimes of the strange particles if the decay proceeded by means of a strong-interaction process. In fact there were some early attempts to do this, based on the suggestion that the decays might be inhibited by a large angular-momentum barrier if the strange particles had high spins. This hypothesis was not particularly successful and was later shown to be false when the spins of these particles were determined, and found to be 0 for the K-mesons and $\frac{1}{2}$ for the Λ, Σ and Ξ.

The outstanding problem therefore was to account for the fact that these particles were *strongly* produced but decayed via the *weak* interaction. It was this behaviour which led to the appellation 'strange' for these particles.

6.2 Strangeness

An explanation of the strong production, weak decay anomaly was offered in 1952 by Gell-Mann and Pais and independently by Nishijima. This explanation proposed that there existed a quantum number associated with the new particles which was conserved in the production process, but not in the decay, which was therefore inhibited relative to the production. It was thus necessary that in the production process the new particles be produced in pairs having opposite values of the new quantum number. Initially it could only be said that the experimental evidence did not contradict the hypothesis of 'associated production', but subsequent evidence demonstrated clearly that this hypothesis was indeed true.

The situation can be illustrated by analogy with the conservation of charge. In the field of the nucleus we can get electron pair production by a gamma ray, that is

$$\gamma \to e^+ e^-,$$

but of course processes like $e^- \to \gamma$ are forbidden by charge conservation. Similarly if we assign values of the new quantum number called strangeness (S) to all particles and if we have the selection rules

for strong interactions strangeness is conserved and $\Delta S = 0$,
for weak interactions strangeness is not conserved and $|\Delta S| = 1$,

then the particles may be produced strongly in pairs, but can decay singly only by means of the weak processes. Strangeness differs from charge in that no force is known to be associated with this quantity. If there is such a force it must be rather weak and masked by other forces.

The next problem is how to assign the strangeness quantum numbers. On a purely empirical basis we may make some deductions from the observations that certain reactions take place. For instance, the process

$$\pi^- p \to \Lambda^0 K^0$$

has been observed in a bubble chamber as a stopping track with two associated Vs. For the π^- and the proton $S = 0$, so that the Λ^0 and K^0 must have strangeness $\pm S$ (where S is greater than or equal to 1) although we cannot say which particle should have positive and which negative strangeness. If we accept the hypothesis that for weak processes $|\Delta S| = 1$, then the decay process $\Lambda \to p\pi^-$ shows that $|S| = 1$ for the Λ^0.

In order to make further progress we must note some other peculiarities of the strange particles. For nucleons and pions the centre of charge of the isotopic-spin multiplet is equal to $\frac{1}{2}B$, where B is the baryon number for the multiplet. Thus for the nucleon doublet the centre of charge is at $\frac{1}{2}(0+1) = +\frac{1}{2}$, for the antinucleons the centre of charge is at $-\frac{1}{2}$ and for the pions at 0, in all cases being given by $\frac{1}{2}B$.

For nucleons and pions the charge of any member of the multiplet is given by the expression

$$Q = I_3 + \frac{1}{2}B,$$

where I_3 is the third component of the isotopic spin.

For the Λ, however, these relationships do not hold. Since all the evidence in familiar processes suggests that the baryon number is conserved we can say that the baryon number for the Λ is 1, since it decays into a proton and a pion. In addition no other particles having the same mass as the Λ, but different charges, have been observed, suggesting that this particle is an isotopic-spin singlet. Thus the centre of charge must be 0, whereas we might have expected it to be $+\frac{1}{2}$. We will find that this displacement of the centre of charge from its 'normal' value is characteristic of the strange particles. If we write $S = -1$ for the Λ, then the centre of charge will be at $\frac{1}{2}(B+S)$. If we replace $\frac{1}{2}B$ by $\frac{1}{2}(B+S)$ in the formula for the charge we obtain

$$Q = I_3 + \tfrac{1}{2}B + \tfrac{1}{2}S. \tag{6.1}$$

In this aspect the strangeness appears to be a measure of the degree to which the centre of charge is displaced from its 'normal' value.

Turning to the Σ-particles, let us first suppose that only the Σ^+ and Σ^- are known, as was the case when the strangeness theory was proposed. From the decay we have $B = 1$ and the observations indicate that the centre of charge is at 0, so that $S = -1$ according to the hypothesis above. If only the charged Σs exist, they would be members of an isotopic-spin doublet having $I_3 = \pm\frac{1}{2}$. However, according to 6.1, this yields charges of $\pm\frac{1}{2}$ so that the scheme would be internally inconsistent. It was proposed in order to account for this difficulty that the Σs in fact formed an i-spin triplet, Σ^+, Σ^0, Σ^-. As before $S = -1$ but now $I = 1$ and $I_3 = +1, 0, -1$ respectively and the charges are $+1, 0$ and -1 as required. The fact that the Σ^0 has open to it an electromagnetic decay mode

$$\Sigma^0 \to \Lambda\gamma,$$

in which $\Delta S = 0$, which is not available to the charged Σs, suggested that it would decay with a lifetime $\sim 10^{-16}$ s, rendering its existence not immediately obvious. As pointed out in the previous chapter, this *prediction* of the Gell-Mann–Nishijima scheme has been amply confirmed by experiments.

Turning to the Ξ-particles we recall that the decay

$$\Xi^- \to \Lambda\pi^-$$

is weak, suggesting $|\Delta S| = 1$ for this process, so that for the Ξ, $S = 0$ or -2. However if $S = 0$ we might expect that the Ξ^- would decay overwhelmingly by the process

$$\Xi^- \to n\pi^-.$$

Since this is not so we conclude that for the Ξ^-, $S = -2$. The centre of charge should therefore be at $-\frac{1}{2}$, implying the existence of a neutral partner for the Ξ^-. The formula 6.1 gives $I_3 = -\frac{1}{2}$ for the Ξ^- and $I_3 = +\frac{1}{2}$ for the proposed Ξ^0. For the Ξ^0 the expected decay is

$$\Xi^0 \to \Lambda\pi^0.$$

As described in the previous chapter this *prediction* of the scheme was also confirmed by experiment.

Note that the Σ^+ and Σ^- are not a particle–antiparticle pair like the π^+ and π^-. For the anti-$\Sigma^+(\bar{\Sigma}^+)$ we expect $Q = -1$, $B = -1$ and $I_3 = -1$ so that $S = +1$, illustrating the general result that particle and antiparticle have opposite strangeness.

We now examine how the Gell-Mann–Nishijima theory applies to the K-mesons. We start from the observation that the process

$$\pi^- p \to \Sigma^- K^+,$$

is observed to occur. Therefore $S = +1$ for the K^+ and using **6.1** we obtain $I_3 = \frac{1}{2}$ for the K^+, consistent with the conservation of I_3. This i-spin assignment implies that the K^+ is one partner of an i-spin doublet, the other partner having $I_3 = -\frac{1}{2}$ and therefore charge 0. This partner is then apparently the K^0.

The question then arises as to what is the status of the K^-. It seems natural that the K^- should be the antiparticle of the K^+ so that it will have $S = -1$. This is the only self-consistent conclusion. To take the K^+, K^0, K^- as an i-spin triplet leads to an internal inconsistency.

The K^- should then form one half of an i-spin doublet, the other partner being a neutral K^0 which is the antiparticle of the K^0 associated with the K^+. The K^0 associated with the K^- is called the $\bar{K}^0$ and has opposite strangeness to the K^0, although these particles are in all other ways identical. The situation is thus slightly different from that of the π^0 which is in *every* way identical with its antiparticle. We will discuss further the peculiar properties of the K^0, $\bar{K}^0$ system in the following chapter.

An interesting consequence of the assignments of strangeness described above concerns the interaction cross-sections for K^+ and K^- mesons. For interactions with nucleons, the baryon number in the initial state is $+1$. Since for $B = +1$ hyperons, Λ and Σ, the strangeness is -1, they can be formed readily even at low energies in $K^- p$ interactions such as

$$K^- p \to \begin{cases} \Lambda^0 \pi^0 \\ \Sigma^0 \pi^0 \\ \Sigma^- \pi^+ \\ \Sigma^+ \pi^-. \end{cases}$$

Reactions like this are not however possible for K^+ mesons. The fact that there are many more reaction channels open for K^- than for K^+ mesons implies that the mean free path for K^- should be very much less than for K^+, as is observed experimentally.

6.3   Hypercharge

A convenient quantity in discussing elementary particles is the hypercharge Y defined as the sum of the baryon number and the strangeness. In terms of the hypercharge we have the relations

Centre of charge $= \frac{1}{2}Y$,

$$Q = I_3 + \tfrac{1}{2}Y.$$

Chapter 7
Spin and parity of the K-mesons and non-conservation of parity in weak interactions

7.1 The τ–θ problem

As the experimental data accumulated it became clear that the masses of the K-mesons decaying into two pions and into three pions were identical, within small error limits. This observation suggested that the two-pion and three-pion decays were alternative decay modes of the same particle. Further weight was lent to this conclusion by the fact that the branching ratio of the two-pion and three-pion modes was found to be independent of the energy of the parent particle and also of its previous history, for instance whether or not it had undergone scattering prior to decay. This implied that the parent particles of the two-pion and three-pion decays, as well as having the same mass, were identical in their interactions with nucleons. Thus, all the evidence suggested that the so-called τ- and θ-particles were in fact different decay manifestations of only one K-meson.

The nature of the puzzle becomes clear only when one considers the possible spin–parity assignments to the τ and θ, and we examine the possible assignments in the following sections.

7.2 Spin–parity for the θ-meson, $K \rightarrow \pi\pi$

Since the pions have zero spin, conservation of angular momentum demands that the spin of the K is equal to the relative orbital angular momentum L of the two pions. The parity of the di-pion is then $(-1)^L$. Thus the possible spin–parity assignments are

$$J^P = 0^+, 1^-, 2^+, 3^-, 4^+, \ldots.$$

In addition if we consider the decay mode of the neutral K-meson into two π^0 mesons it is clear that since we are dealing with bosons, and since the space part of the wave function is symmetrical the Pauli principle allows only symmetrical angular momentum wave functions so that $J^P = 1^-, 3^-, \ldots$ are not allowed. Thus, assuming that the charged and neutral K-mesons decaying into two pions are different members of an i-spin multiplet and have the same spin–parity, then if parity is conserved in the decay the only possible J^P assignments are those of even spin and even parity.

7.3 Spin–parity for the τ-meson, K → 3π

One can treat the three-pion system most readily in terms of a di-pion, which in the case of the charged decay we take to consist of the two pions of like charge, plus an added third pion. We take the relative orbital angular momentum in the

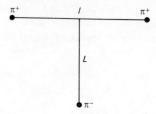

Figure 49 Definition of the angular momenta in the charged three-pion system

di-pion to be l, and of the third pion *relative* to the di-pion to be L (see Figure 49). The parity of the three-pion system is then

$$(-1)^3(-1)^l(-1)^L = -(-1)^L$$

since the di-pion must have even l for symmetry reasons.

The spin J of the three-pion system must lie in the interval

$$|L-l| \leqslant J \leqslant |L+l|.$$

The possible spin–parity assignments are then summarized in Table 3.

Table 3

l	L	J^P
0	0	0^-
0	1	1^+
0	2	2^-
2	0	2^-
2	1	$1^+, 2^+, 3^+$
2	2	$0^-, 1^-, 2^-, 3^-, 4^-$

Comparing those assignments with those for the θ-meson we see that the first possible J^P value common to both θ and τ is 2^+ for $l = 2$, $L = 1$. Higher angular-momentum values could also give compatible J^P assignments, but we neglect them since decays from such states will be seriously inhibited by the high angular-momentum barriers. Indeed, even for the state mentioned, we might expect that the three-pion decay mode would be markedly inhibited for this reason.

The further study of this problem requires the determination of the spin–parity from an examination of the decay angular distributions for the meson. In order to see what form of angular distribution is to be expected for different J^P assignments we write the angular part of the final three-pion wave function in the c.m.s. as a sum of partial waves specified by l and L. In order to examine the final state we define the following quantities;

(a) q is the relative linear momentum of the pions in the di-pion.

(b) p is the linear momentum of the third pion relative to the di-pion.

(c) θ is the angle between $\mathbf{p}$ and $\mathbf{q}$ so that $\cos \theta = (\mathbf{p} \cdot \mathbf{q})/pq$

The relative proportions of the different angular momentum states contributing to the final state will be fixed by the Clebsch–Gordan coefficients (see Appendix B) but may also depend on the relative energies (proportional to p^2 and q^2) in the three-pion system. We specify this energy-dependent weighting factor by $f_{L,l}(p^2, q^2)$. In the general case we can then write our final-state wave function as

$$\sum_{L, l, m_L, m_l} C(L, l, J; m_L, m_l)\, Y_L^{m_L}(\theta_p, \phi_p)\, Y_l^{m_l}(\theta_q, \phi_q)\, f_{L,l}(p^2, q^2),$$

where with our coordinate system this reduces to

$$\sum_{L, l, m_l} C(L, l, J; 0, m_l)\, Y_l^{m_l}(\theta)\, f_{L,l}(p^2, q^2), \tag{7.1}$$

since $\cos \theta_p = 1$, $m_L = 0$ and $Y_L^{m_L} = $ a constant.

It remains to discuss the form of the function $f(p^2, q^2)$. In practice we resort to an approximation by assuming that the asymptotic solutions of the Schrödinger equation are valid. This is equivalent to the assumption that any momentum dependence arising from the internal structure of the τ-meson may be neglected. The approximation is good if the pion wavelengths are long compared with the τ-radius, so that there is a high probability that the pions lie 'outside' the τ (compare the deuteron). With this approximation we can write

$$f = c_{L,l} p^L q^l, \tag{7.2}$$

where $c_{L,l}$ is a constant.

In order to obtain manageable (and realistic) formulae we also make the approximation of retaining only the lowest angular-momentum values consistent with any given spin and parity, using the fact that the angular-momentum barrier will inhibit the higher terms in comparison with the lower ones.

With these approximations we can write down the values of the square of the modulus of the matrix element for the decay process using 7.1 and 7.2. These quantities are given in Table 4.

Table 4

J	P	L	l	$\|M\|^2 \propto$	$I(\varepsilon)$ not including phase space
0	+	not possible			
0	−	0	0	1	1
1	+	1	0	p^2	ε
1	−	2	2	$p^4 q^4 \sin^2\theta \cos^2\theta$	$\varepsilon^2(1-\varepsilon)^2$
2	+	1	2	$p^2 q^4 \sin^2\theta$	$\varepsilon(1-\varepsilon)$
2	−	$\begin{cases} 2 \\ 0 \end{cases}$	$\begin{cases} 0 \\ 2 \end{cases}$	depends on mixing	
3	+	$\begin{cases} 3 \\ 1 \end{cases}$	$\begin{cases} 0 \\ 2 \end{cases}$	depends on mixing	
3	−	2	2	$p^4 q^4 \sin^2\theta(5+3\cos^2\theta)$	$\varepsilon^2(1-\varepsilon)^2$

In order to obtain the final energy or momentum distributions we must multiply $|M|^2$ by the appropriate phase-space factor. For the angular distributions the transition probability is simply given by

$$\frac{dN(\theta)}{d(\cos\theta)} \propto I(\theta),$$

where $I(\theta)$ is the angular function in $|M|^2$, as given in the table.

It is convenient to parameterize the energy dependence of the decay in terms of $\varepsilon = E(\pi^-)/E_{max}$ where E_{max} is the maximum possible pion energy and is equal to two-thirds of the available c.m.s. energy. We may then write the phase-space factor as

$$\varepsilon^{\frac{1}{2}}(1-\varepsilon)^{\frac{1}{2}}\, d\varepsilon\, d(\cos\theta),$$

and expressing the pq dependence in terms of ε, as given in the table, we obtain for the energy distribution

$$\frac{dN(\varepsilon)}{d\varepsilon} = I(\varepsilon)\, \varepsilon^{\frac{1}{2}}(1-\varepsilon)^{\frac{1}{2}},$$

where $I(\varepsilon)$ is the energy-dependent part of the matrix element, given in the table. Experimentally the distribution in $\cos\theta$ is found to be nearly isotropic (see Figure 50) favouring J^P equal to 0^- or 1^+, but not 1^-. The best fit to the distribution in ε leads to $I(\varepsilon)$ being constant (that is the energy distribution after extraction of the phase-space factor). This distribution thus indicates that J^P is 0^-. The conclusion of the analysis is thus that, assuming the τ and the θ to be the same particle, parity cannot be conserved in the decay of this K-meson.

It is important to bear in mind that this analysis depends on certain assumptions.

(a) Validity of the 'small radius' approximation used to obtain the form of f.

(b) The treatment has been non-relativistic, although any errors due to this approximation are likely to be small in view of the small Q-value of the decay.

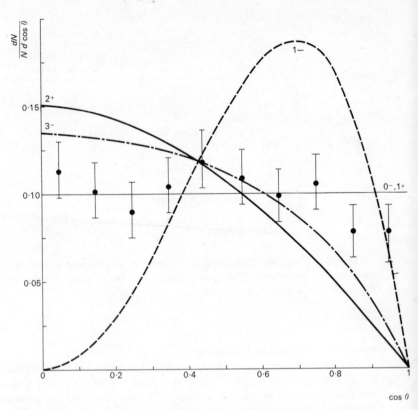

Figure 50 The distribution of cos θ (see text for definition of θ) for τ-decay.
The curves show the calculated distributions for various spin–parity assignments.
The approximately isotropic distribution agrees only with J^P = 0⁻ or 1⁺
(Baldo-Ceolin *et al.*, 1957)

(c) We have neglected the effects of any two-pion interactions in the final state. Other data on the two-pion system suggest that this assumption is justified at the energies in question.

(d) We have neglected the possibility that the observed distributions might be due to the mixing of, for instance, $L = 3, l = 0$ and $L = 1, l = 2$ states. That the mixing should be such as to produce distributions simulating the $L = 0, l = 0$ state seems sufficiently improbable that it may be neglected.

7.4   Distribution in the Dalitz plot for the τ-decay

It is shown in Appendix A.7 that the Dalitz plot has the property that, if the distribution of the energies in a three-particle system is determined only by the phase-space factor, then the Dalitz plot will be uniformly populated. It is of interest to investigate how the matrix-element dependence on p, q and θ, as discussed in the previous section, influences the distribution in the plot. We use the form of the plot where for each event the perpendicular distances from the point representing that event to the sides of the triangle are proportional to the kinetic energies of the mesons (see Figure 51). Thus PN is proportional to the kinetic energy of the 'unlike'

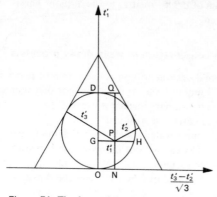

Figure 51 The form of the Dalitz plot for three-pion decay.
t'_1, t'_2, t'_3 are the c.m.s. kinetic energies of the pions.
The circle corresponds to the classical limit for momentum conservation

meson, so that PN $\propto p^2$. It may then be shown that PQ $\propto q^2$ and $\cos \theta \propto$ GP/GH. We recall that momentum conservation limits the points to within a circle inscribed in the triangle (non-relativistic case). An angular correlation between p and q is reflected by a variation in the density of points across the circle at any fixed value of PN. Points of constant $\cos \theta$ lie on an ellipse having major axis DO. On DO, $\theta = \frac{1}{2}\pi$. On the circumference of the circle $\theta = 0$ so that the pions are collinear, a general result for the boundary of any form of Dalitz plot. Variations of density will also occur due to the terms dependent on p and q so that, for instance, the average density along lines parallel to the base of the triangle might be expected to vary with the Lth power of PN.

Special arguments concerning the distribution of points in the Dalitz plot may be made for limiting energies and angles. These arguments do not depend on the approximations made in the fuller treatment. For instance;

(a) A pion of zero energy is not possible if the spin–parity of the parent particle is odd–odd or even–even. Such a pion would need to be in an s-state, since otherwise it could not penetrate the centrifugal barrier. Thus for an odd-parity parent the remaining pions must be in a state of even angular momentum, resulting in an even value for the total spin. Similarly for an even-parity parent the remaining pions must be in a state of odd angular momentum, giving total spin odd. Thus for J^P odd–odd or even–even, the density of points in the Dalitz plot should decrease towards the region corresponding to small pion momenta, i.e. the regions where the circle touches the triangle.

(b) A pion of unlike charge, with maximum energy, is not allowed if the spin–parity of the parent particle is odd–odd or even–even. In such a case, for instance, the two positive pions would have zero relative momentum so that the total spin will be the orbital angular momentum of the π^- relative to the di-pion. This is seen to exclude J^P odd–odd or even–even.

7.5 Other examples of parity non-conservation in weak decay processes

The explanation of the τ–θ anomaly as due to non-conservation of parity in the decay was proposed by Lee and Yang in 1956. At the same time they pointed out that no test of parity conservation in ordinary nuclear β-decay had ever been made and they suggested experiments to carry out such tests.

The critical experiment which demonstrated conclusively that in fact parity was *not* conserved in nuclear β-decay was carried out immediately thereafter by Wu and collaborators (1957). In order to understand the principle underlying this experiment we note that the parity transformation reverses the direction of any

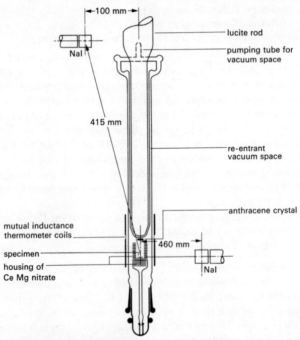

Figure 52 The cryostat and counters used by Wu *et al.* (1957) in the detection of parity violation in ^{60}Co decay

momentum vector **p** but does not reverse the direction of angular-momentum vectors **L**, since **L** is formed by a product of momentum and position vectors *both* of which change sign under the parity transformation. In the experiment of Wu, Ambler, Hayward, Hoppes and Hudson the distribution of β-particles emitted by an aligned radioactive source was examined. Consider electrons emitted in the same direction as the nuclear spin, or even into the hemisphere around the spin orientation. Applying the parity transformation to this situation the electron direction reverses, the spin direction remains unchanged and the electrons are now emitted opposite to the spin. Parity will only be conserved if the distribution of emitted electrons is symmetrical with respect to the spin orientation.

The apparatus used by Wu *et al.* is shown in Figure 52. The source used was ^{60}Co which emits electrons of 0·312 MeV followed by a cascade of two gamma rays of 1·19 and 1·32 MeV to give ^{60}Ni. In order to align the ^{60}Co nuclei the source was deposited in a crystal of cerium magnesium nitrate, which exerts a very strong internal magnetic field, and the thermal motions which would destroy the alignment were reduced to a minimum by cooling the crystal to 0·01 K by adiabatic demagnetization. The ^{60}Co nuclei were aligned by applying an external magnetic field and the degree of alignment was measured by observing the anisotropy of the emitted gamma rays, which were detected by the sodium iodide crystals. When the cooling was complete, the magnetic-field coils were removed and the decay electrons counted by means of an anthracene scintillator placed above the sample. The counting rate was measured as a function of time for nuclei with their spins aligned upwards and downwards. The results are shown in Figure 53 and show a

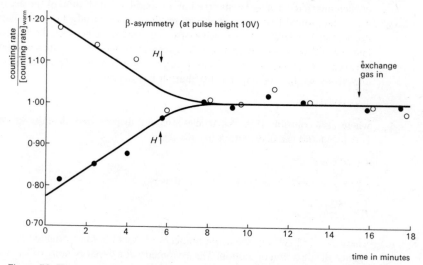

Figure 53 The asymmetry in the β-ray counting rate for the aligned ^{60}Co nuclei in the experiment of Wu *et al.* (1957). A clear correlation between the β-direction and the alignment is observed, which decreases as the sample warms up and the alignment disappears

clear anisotropy or correlation between the spin direction and the direction of emission, which decreases to zero as the sample warms up and the alignment disappears. Following our preliminary discussion, this observation indicates clear parity violation in the β-decay. This experiment is an example of a general method of detecting the effects of non-conservation of parity by looking for the effects on any process of terms, in the Hamiltonian for the system, which are not invariant under the parity transformation. Any pseudo-scalar term can produce such evidence of parity non-conservation. If **L** represents the spin vector for the aligned nuclei then the quantity $\langle \mathbf{L} \rangle . \mathbf{p}$ is a pseudo-scalar, where $\langle \mathbf{L} \rangle$ is the expectation value of **L**. If parity is not conserved then the intensity of the electrons may be a function of this quantity, leading to an electron distribution of the form

$$1 + x \cos \theta,$$

where we define $\cos \theta$ as $\mathbf{L}.\mathbf{p}/|\mathbf{L}.\mathbf{p}|$ and x is a constant.

Among other examples of parity non-conservation we may study the angular distribution of the Λ-decay. Taking the Λ-spin to be $\frac{1}{2}$ we see that the relative angular momentum of its decay products, the pion and the nucleon, can be 0 or 1. The states of zero or unit relative angular momentum will have odd or even parity respectively. If parity is conserved in the decay process only one of these states can be present, while if parity is not conserved we may have both states present and we may get interference between them. If we have Λ-particles which are polarized preferentially, either along or against the direction of their motion, then the effect of such interference may be observed in the angular distribution of the decay and it is in fact found that many processes resulting in the production of Λ-particles do yield such polarized Λs. In such a case, if θ is the angle between the Λ-direction and one of the decay particles in the decay centre-of-mass system, then interference between S-wave and P-wave decay processes will lead to a forward–backward anisotropy so that the angular distribution has the form

$$1 + \alpha \cos \theta,$$

where α is a constant. If the Λ-particles are not fully polarized, then the anisotropy is reduced and the distribution has the form

$$1 + \alpha P \cos \theta,$$

where P is the polarization of the Λs, that is

$$P = \frac{N^+ - N^-}{N^+ + N^-},$$

where N^+ and N^- represent the numbers of Λs with spins oriented along and against the direction of motion. The asymmetry due to parity non-conservation in the decay has been observed in a number of experiments.

A fuller analysis shows that the proton emitted in the Λ-decay should be polarized if parity is not conserved. Such a polarization has been observed. It implies that the decay is a function of the pseudo-scalar formed by the product of the proton spin and momentum.

A further consequence of parity non-conservation in β-decay processes involving neutrinos is that, for β-particles emitted even from non-aligned nuclei, the electrons are polarized such that their spin lies in the same direction as the motion. This is sometimes expressed by saying that the helicity, which we may define as the cosine of the angle between the spin and the direction of motion, is equal to $+1$.

Similar tests have demonstrated that parity is not conserved in the decay of the π- and μ-mesons and indeed in no weak interaction process which has been investigated has parity conservation been found to hold good.

Chapter 8
Invariance under the *CP* and *T* operations, properties of K⁰ mesons and other selection rules for weak decays

8.1 The *TCP* theorem

The *TCP* theorem, derived in different forms by several workers (Schwinger in 1953, Luders in 1954 and Pauli in 1955), states that, for locally-interacting fields, a Lagrangian which is invariant under proper Lorentz transformations is invariant with respect to the combined operation *TCP*. By the operation *TCP* is meant the set of operations time reversal, charge conjugation (i.e. particle–antiparticle exchange) and the parity transformation, taken in any order. The proof applies only to the combined set of operations even though the theory may not be invariant under the individual operations *T*, *C* and *P*.

It is an obvious consequence of the *TCP* theorem that if an interaction is not invariant under one of the operations it must also fail to be invariant under one or other of the remaining two. Equally if a process is invariant under one of the operations, it must also be invariant under the product of the other two.

We have already seen that for weak decays parity is not conserved, so that if these processes are to be invariant under time reversal they cannot be invariant under charge conjugation. A number of tests have demonstrated that this is in fact the case. For instance, in the decay of muons we have already mentioned that the decay electrons are fully polarized. If charge conjugation holds, we should expect that the helicity of the positron from μ^+ decay should be the same as that of the electron from μ^- decay, since these two processes are the charge conjugates of each other and since the only effect of the charge conjugation operator is to interchange particle and antiparticle. The helicities of the electrons and positrons from muon decay have been measured by observing the transmission through magnetized iron of the bremsstrahlung photons emitted by these particles. The transmission coefficient is a function of the photon helicity, which depends on the helicity of its parent particle. The result of these measurements indicated that in fact the positrons were fully right-handed, and the electrons fully left-handed, giving maximum violation of charge conjugation. Similar effects have been observed for both pion decay and nuclear β-decay.

A further interesting result concerning charge conjugation and parity can be derived with the aid of the TCP theorem. It can be shown that if the Hamiltonian responsible for a transition is invariant under charge conjugation, then the parity conserving and parity non-conserving final states cannot interfere with each other. Thus the observation of such interference effects, as for instance in the decay of the Λ-particle, demonstrated that not only is parity not conserved in this process, but neither is the process invariant under charge conjugation.

8.2 The K^0 particles and invariance under the operator CP

In order to discuss this problem we must consider the nature of the state functions for the K^0 and its antiparticle the $\bar{K}^0$. On the basis of our earlier discussion in Chapter 7, we assume spin zero for the K^0s. In fact, for bosons which are different from their antiparticles the state function must be complex. Although this fact can only be proved by reference to the field theory for these particles, it may be rendered more plausible by reference to the modifications necessary to the Schrödinger equation for the inclusion of an electromagnetic field. In such a case we must replace the operators $\partial/\partial x$ by $(\partial/\partial x) - ieA$ and $\partial/\partial t$ by $(\partial/\partial t) + ieV$ where A and V are the vector and scalar potentials respectively. A consequence of this modification is that the wave functions representing the field must be complex. We therefore write the wave functions in the form

$$\phi = \phi_1 + i\phi_2,$$

where ϕ_1 and ϕ_2 are real. In addition the continuity equations require that particles represented by complex conjugate wave functions have opposite charge and current densities. Thus the antiparticle of that represented by the wave function ϕ will be represented by

$$\phi^* = \phi_1 - i\phi_2.$$

We may now use such wave functions for the kaon field and examine the effects of the charge conjugation operator. We know that

$$C\phi = \phi^*$$

so that $\quad C\phi_1 = \phi_1 \quad$ and $\quad C\phi_2 = -\phi_2.$

For particles such as the π^0 and the photon, which are the same as their anti-particles, $\phi_2 = 0$. Under charge conjugation, however, the wave function may still change sign since the eigenvalue of C may be ± 1. For charged particles and neutral particles which have distinct antiparticles, ϕ_2 is not equal to zero. Experiment shows that the K^0 and the $\bar{K}^0$ are certainly different, having opposite strangeness, so that they should be represented by complex wave functions.

We note however that there is an important difference between the situation for the neutral K-mesons and that for other particle–antiparticle pairs. For charged particles, virtual transitions between particle and antiparticle are always forbidden by charge conservation. In addition, for baryons, such as for instance the neutron and the antineutron, virtual transitions between particle and antiparticle are forbidden by baryon conservation. The laws that forbid these transitions are true for all kinds of interactions. For the K^0 and the $\overline{K}^0$, on the other hand, virtual transitions between particle and antiparticle are possible via the weak interaction since they involve only violation of strangeness conservation. We may thus expect to find *mixing* of the neutral K-mesons, and this effect was predicted by Gell-Mann and Pais in 1955.

We write

$$\phi_{K^0} = \frac{1}{\sqrt{2}}(\phi_1 + i\phi_2)$$

and $\quad \phi_{\overline{K}^0} = \frac{1}{\sqrt{2}}(\phi_1 - i\phi_2).$

Although we do not expect invariance of the wave functions of the decaying particles under C and P separately, we do expect invariance under the operation CP. Applying this operator to the wave functions ϕ_{K^0} and $\phi_{\overline{K}^0}$ we can adjust the relative phase of the states to get

$$CP\phi_{K^0} = \phi_{\overline{K}^0} \quad \text{and} \quad CP\phi_{\overline{K}^0} = \phi_{K^0}.$$

Thus ϕ_{K^0} and $\phi_{\overline{K}^0}$ are not eigenfunctions of CP.

For the decay process we do not require eigenfunctions of the strangeness operator, and Gell-Mann and Pais proposed for the reasons described above that the appropriate wave functions should be the mixed wave functions ϕ_1 and ϕ_2

$$\phi_1 = \frac{1}{\sqrt{2}}(\phi_{K^0} + \phi_{\overline{K}^0}) \quad \text{and} \quad \phi_2 = -\frac{i}{\sqrt{2}}(\phi_{K^0} - \phi_{\overline{K}^0}).$$

These functions are indeed eigenfunctions of CP since

$$CP\phi_1 = \frac{1}{\sqrt{2}}(CP\phi_{K^0} + CP\phi_{\overline{K}^0}) = \frac{1}{\sqrt{2}}(\phi_{\overline{K}^0} + \phi_{K^0}) = \phi_1$$

and $\quad CP\phi_2 = \frac{i}{\sqrt{2}}(\phi_{\overline{K}^0} - \phi_{K^0}) = -\phi_2.$

The eigenvalues of ϕ_1 and ϕ_2 under CP are thus $+1$ and -1 respectively. These are the functions appropriate for the K^0 decays and the particles corresponding to which we might expect to exhibit different decay features. They are known as the K_1^0 and the K_2^0.

We may now examine the effect of the CP operation on the final state in the decay. For a two-pion system of relative orbital angular momentum L we have seen that the parity operator has eigenvalue $(-1)^L$. Application of the charge-conjugation operator interchanges the positive and negative pions in the charged decay. As shown in Scheme 1 (using the usual property of the spherical harmonics) the eigenvalue of C is also $(-1)^L$. Thus the eigenvalue of the combined operation PC is $(-1)^{2L}$; i.e. the two-pion system has eigenvalue $+1$ under this operation. This means that to preserve invariance under the CP operation *only* the K_1^0 can decay to two pions.

Scheme 1 Effect of the CP operation on a $\pi^+\pi^-$ system with relative orbital angular momentum L

$$CP\left(\overbrace{\pi^+ \,\uparrow^L\, \pi^-}\right) \rightarrow C(-1)^L\left(\overbrace{\pi^+ \,\uparrow^L\, \pi^-}\right)$$

$$\rightarrow (-1)^L\left(\overbrace{\pi^- \,\uparrow^L\, \pi^+}\right)$$

$$\rightarrow (-1)^L\left(\overbrace{\pi^+ \qquad \pi^-}_{\downarrow_L}\right)$$

$$\rightarrow (-1)^L(-1)^L\left(\overbrace{\pi^+ \,\uparrow^L\, \pi^-}\right)$$

thus $\quad CP[\psi(\pi^+\pi^-)] = (-1)^{2L}\psi(\pi^+\pi^-)$

$$= +1\,\psi(\pi^+\pi^-).$$

In a similar way it may be shown that the eigenvalues for a three-pion system under the CP operation are $-(-1)^L$ where L is the orbital angular-momentum quantum number of the neutral pion with respect to the di-pion (Scheme 2). For the K_2^0 with eigenvalue -1 under CP only the three-pion or one-pion and two-lepton modes of decay are permitted. Gell-Mann and Pais proposed that the lifetime for these three-particle decay modes, that is for the K_2^0, should be about a

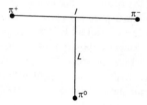

Figure 54

Scheme 2 The effect of the CP operation on a $\pi^+\pi^-\pi^0$ system shown in Figure 54

$CP[\psi(\pi^+\pi^-\pi^0)]$

$\rightarrow (-1)^3(-1)^l(-1)^L C[\psi(\pi^+\pi^-\pi^0)]$

$\rightarrow -(-1)^{l+L} C[\phi(\pi^+\pi^-)]\alpha(\pi^0)$

$\rightarrow -(-1)^{l+L}(-1)^l \psi(\pi^+\pi^-\pi^0)$

$\rightarrow -(-1)^L \psi(\pi^+\pi^-\pi^0).$

thousand times longer than for the K_1^0 on the basis of the decay rate for the similar process

$$K^+ \to \mu^+ \pi^0 \nu,$$

for the charged K-meson.

We should expect that the three-particle modes be inhibited relative to the two-particle decays simply due to the difference in available phase space, and also due to the angular-momentum barrier. Thus the K_1^0 will not in general decay via the three-particle modes. For the $(\pi^0 \pi^0 \pi^0)$ combination the eigenvalue under PC can only be -1, so that K_1^0 decay to three neutral pions is forbidden by PC invariance.

8.3 The development of a K^0 beam

We shall see in the next section that even the postulate of invariance under the CP operation has turned out to be not exactly true. However the violation is very small and the conclusions which we shall draw here, based on the results of the last section, are largely unaffected.

We shall consider the development of a beam of K^0 mesons. Let us suppose that the K^0s are generated in the process

$$\pi^- p \to \Lambda^0 K^0.$$

In this strong interaction, strangeness is conserved and we have a pure K^0 state which we may regard as consisting of 50 per cent of K_1^0s and 50 per cent of K_2^0s. After about 10^{-9} s, as measured in the kaon rest system, nearly all the K_1^0s will have decayed into pion pairs. The intensity of the beam has been reduced to half the original intensity and it now consists of almost pure K_2^0 particles. A xenon bubble chamber has been used to show that 0.53 ± 0.05 of all K^0s decay via two-pion modes. This remaining beam of K_2^0s is no longer a state of pure strangeness, having equal strangeness $+1$ and strangeness -1 components. Thus, whereas the original K^0 beam could produce only reactions having a final state of strangeness $+1$, the stale beam can produce reactions of final-state strangeness -1, such as

$$\bar{K}^0 n \to \Sigma^- \pi^+,$$

$$\bar{K}^0 p \to \Lambda \pi^+,$$

as well as the $S = +1$ reactions.

We recall that the cross-section for reactions involving strangeness -1 particles is much greater than that for strangeness $+1$, so that if the stale beam is allowed to pass through material then the $S = -1$ component is preferentially removed. Thus this process will again produce a preponderance of K^0 over $\bar{K}^0$ particles, the beam therefore again containing a proportion of K_1^0 which will decay by two-pion modes.

Experiments have in fact demonstrated this *regeneration* property and have shown that a beam consisting initially only of K^0 mesons can be used after a suitable time to produce Λ- and Σ-particles as predicted by the Gell-Mann scheme.

Several authors have suggested an illustrative analogy between the behaviour of the K^0s and that of polarized light.

Consider a circularly polarized beam of light. This may be regarded as consisting of beams plane-polarized in orthogonal directions, which may be thought of as corresponding to the K_1^0 and K_2^0. If the beam is now passed into a birefringent material, one plane-polarized component will be removed leaving plane-polarized light. We take this as the analogue of the decay of the K_2^0. If the surviving component is now passed into a quarter-wave quartz plate then the plane of polarization will be rotated and once more we have components in both the original planes of polarization, a process which we compare to the regeneration of the K_1^0.

8.4 Time variation of the neutral kaons

We recall that for a stable stationary-state solution of the wave equation, for a particle of mass m, the solution contains a phase factor e^{-imt}, where as usual we have taken units such that $\hbar = c = 1$ and m is in units of reciprocal time.

If the state is unstable, undergoing an exponential decay, then an additional phase factor $e^{-(\frac{1}{2})\Gamma t}$ must be included, where Γ is the decay width, so that the mean lifetime is equal to Γ^{-1}. The combined phase-factor is often written as e^{-iMt}, where

$$M = m - \tfrac{1}{2}i\Gamma.$$

In these terms we can write the non-relativistic wave function for the K^0 state, after a proper time t, as

$$\psi(t) = \frac{1}{\sqrt{2}}\left(|K_1^0\rangle e^{-iM_1 t} + i|K_2^0\rangle e^{-iM_2 t}\right), \qquad \textbf{8.1}$$

where $M_1 = m_1 - \tfrac{1}{2}i\Gamma_1$ and $M_2 = m_2 - \tfrac{1}{2}i\Gamma_2$ and m_1 and m_2 are the K_1^0 and K_2^0 masses. As required, at time $t = 0$

$$\psi(0) = \frac{1}{\sqrt{2}}\left(|K_1^0\rangle + i|K_2^0\rangle\right).$$

It is clear from the expression **8.1** that $K_1^0 - K_2^0$ interference is possible and that the interference terms will involve the mass difference, $\Delta m = m_2 - m_1$, between the K_1^0 and the K_2^0. As time passes the K_1^0 component decreases relative to the K_2^0 component, because $\Gamma_1 \gg \Gamma_2$, and in addition its phase changes relative to the K_2^0 by an amount which is a function of Δm. This result is the key to a number of different interference effects and affords a method of measuring Δm.

Possibly the most obvious such interference effect is that of 'strangeness oscillation'. Let us evaluate the intensity $|\psi(t)|^2$, where we write $\psi(t)$ explicitly in terms of $|K^0\rangle$ and $|\bar{K}^0\rangle$

$$\psi(t) = \tfrac{1}{2}(|K^0\rangle + |\bar{K}^0\rangle)e^{-iM_1 t} + \tfrac{1}{2}(|K^0\rangle - |\bar{K}^0\rangle)e^{-iM_2 t}.$$

Then multiplying by the complex conjugate quantity, and extracting the appropriate terms, we find for the intensity of the K^0 component

$$N(\mathrm{K}^0) \propto \tfrac{1}{4}[e^{-\Gamma_1 t}+e^{-\Gamma_2 t}+2\cos[(m_2-m_1)]\,e^{-\frac{1}{2}(\Gamma_1+\Gamma_2)t}], \qquad \textbf{8.2}$$

and for the $\bar{\mathrm{K}}^0$

$$N(\bar{\mathrm{K}}^0) \propto \tfrac{1}{4}[e^{-\Gamma_1 t}+e^{-\Gamma_2 t}-2\cos[(m_2-m_1)t]\,e^{-\frac{1}{2}(\Gamma_1+\Gamma_2)t}]. \qquad \textbf{8.3}$$

As expected

$$N(\mathrm{K}^0)+N(\bar{\mathrm{K}}^0) \propto \tfrac{1}{2}(e^{-\Gamma_1 t}+e^{-\Gamma_2 t}).$$

For times short compared with $\tau_2 (= 1/\Gamma_2)$, we have the simpler expressions

$$N(\mathrm{K}^0) \propto \tfrac{1}{4}[1+e^{-\Gamma_1 t}+2\cos(\Delta m t)\,e^{-\frac{1}{2}\Gamma_1 t}]$$
$$N(\bar{\mathrm{K}}^0) \propto \tfrac{1}{4}[1+e^{-\Gamma_1 t}-2\cos(\Delta m t)\,e^{-\frac{1}{2}\Gamma_1 t}].$$

Thus in principle a study of the variations with time of the number of K^0 and $\bar{\mathrm{K}}^0$ mesons in a beam consisting originally purely of K^0 particles affords a method of determining the mass difference Δm. Some examples of the effects to be expected for different values of Δm are shown in Figure 55. This figure illustrates the extraordinary sensitivity of the method, which makes it possible to measure a mass difference of as little as 10^{-5} eV or 10^{-38} g between the two K^0s.

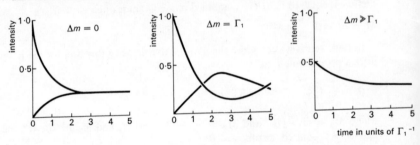

Figure 55 Relative intensities of K^0 and $\bar{\mathrm{K}}^0$ as a function of time, in a beam starting as pure K^0, for different values of Δm

In practice it is possible to produce K^0 mesons in a bubble chamber and to study the proportion of K^0 as a function of time, identifying such $S = -1$ particles by their production of hyperons. Several experiments of this kind have been performed yielding values for $\Delta m/\Gamma_1$ between $0\cdot62^{+0\cdot33}_{-0\cdot27}$ and $1\cdot9\pm0\cdot33$.

Even this crude measure of the value of Δm may be used to infer that $\Delta S = 2$ transitions are unimportant in weak interactions, since it may be shown that such transitions require values of $\Delta m \sim 10^{-17}\,\mathrm{s}^{-1}$, differing from the observed value by about seven orders of magnitude.

8.5 Failure of invariance under CP in K^0 decay

Having analysed the K^0 decay phenomena on the basis of invariance under CP, we must now discuss the evidence provided by more recent experiments which.

shows quite clearly that there is a violation of *CP* invariance in these processes. The analysis in the preceding section is nevertheless a very good approximation to reality, since the degree of *CP* violation is very small. We may note that this is in sharp contrast to the violation of parity invariance in weak interactions, where the breaking is maximal.

The first experiment which demonstrated failure of *CP* invariance was carried out by Christenson, Cronin, Fitch and Turlay (1964), who detected a two-pion decay mode for the K_2^0 while studying regeneration phenomena.

The detection apparatus is shown in Figure 56. The experiment was carried out

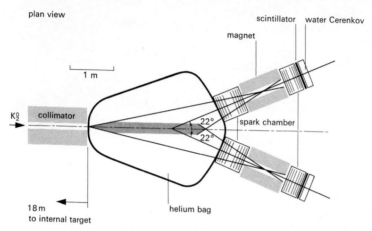

Figure 56 Arrangement used by Christenson, Cronin, Fitch and Turlay (1964) for detection of the two-pion decay of the K^0 meson. The region seen by the two detection arms is heavily shaded

at the Brookhaven proton synchrotron where a beam from an internal target was selected at 30° to the direction of the 30 GeV internal circulating proton beam by means of a lead collimator. On the target side of the collimator was placed a 4 cm thick block of lead which acted as a filter for γ-rays, while a bending magnet after the collimator swept charged particles from the beam. The detecting apparatus was placed behind a further collimator at a distance of 18 m from the target at which point only K_2^0 mesons and neutrons remain in the beam. Some γ-rays may also be present, although attenuated by the lead before the first collimator. The detecting apparatus consisted of a helium-filled bag, particles from which were detected by a pair of symmetrically placed spark-chamber spectrometers. Each of these spectrometers consisted of a pair of spark chambers, separated by a magnet, and triggered by scintillation and Cerenkov counters, positioned as shown in the figure. The chambers are triggered only when a beam particle decays into charged particles of velocity $\gtrsim 0.75c$ passing into the two spark-chamber assemblies.

$K^0 \rightarrow 2\pi$ decays were unambiguously detected in the following way. When two particles of opposite charge were detected in coincidence in the spark chambers the resultant momentum of the pair of particles was calculated, and also their effective mass, on the assumption that the particles were pions. The effective mass is given by

$$M_{1,2}^2 = (E_1 + E_2)^2 - (\mathbf{p}_1 + \mathbf{p}_2)^2$$
$$= M_1^2 + M_2^2 + 2(M_1^2 + p_1^2)^{\frac{1}{2}}(M_2^2 + p_2^2)^{\frac{1}{2}} - 2p_1 p_2.$$

The effective mass could then be studied as a function of the angle between the resultant two-particle momentum and the tightly collimated K_2^0 beam. Alternatively the distribution in this angle could be studied as a function of the effective mass, as is shown in Figure 57 for three regions of effective two-particle mass. These

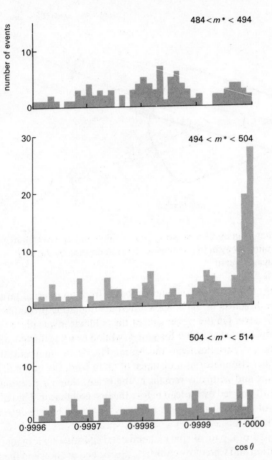

Figure 57 Angular distribution near the forward direction (note scale) of the resultant two-particle momentum, for three ranges of two-particle effective mass m^*, in the decay of the long-lived K^0 (Christenson et al., 1964)

plots show quite clearly the existence of a long-lived K^0 meson *which decays into two pions*. A three-particle decay would not show this sharp correlation between K^0 mass and angle, while the density of helium in the apparatus was quite inadequate to give sufficient regeneration to explain the number of events observed.

Christenson, Cronin, Fitch and Turlay obtained

$$R = \frac{K_2^0 \to \pi^+\pi^-}{K_2^0 \to \text{all charged}} = (2 \cdot 0 \pm 0 \cdot 4) \times 10^{-3},$$

and this result has been confirmed in a number of subsequent experiments using K_2^0 particles having a wide range of momenta.

As is clear from our earlier discussion, the presence of such a two-pion decay mode implies that the K_2^0 cannot be a pure eigenstate of the operator CP. It is natural to ask to what extent the all-neutral decay

$$K^0 \to \pi^0\pi^0$$

also occurs for the long-lived kaon. Although such measurements are more difficult than for the charged system, values have been obtained. In an experiment at CERN all four γ-rays from the π^0s were detected by means of pair-production in thick-plate spark chambers, with absolute calibration by comparison with the event rate from the K_1^0 mesons in a carbon target placed in the apparatus for this purpose. One difficulty of this method is to correct for events of the kind $K^0 \to 3\pi^0 \to 6\gamma$, where two γ-rays do not produce pairs and the remaining four simulate the $K^0 \to 2\pi^0$ decay. Such a background was corrected by means of a 'Monte Carlo' computer calculation, where fake events of the kind $K^0 \to 3\pi^0 \to 6\gamma$ were tested to determine in how many cases they might simulate a $K^0 \to 2\pi^0$ decay.

An experiment carried out at Princeton used a different technique, in which only one of the four γs was detected. It distinguishes the $K^0 \to 2\pi^0$ decay from all other modes, such as $K^0 \to 3\pi^0$, $K^0 \to \pi^+\pi^-\pi^0$, $K^0 \to \pi^0\pi^0\gamma$, $K^0 \to \pi^+\pi^-\gamma$, by exploiting the fact that only in this mode are γs produced with an energy greater than 170 MeV, in the K^0 centre-of-mass system. The energies of the γ-rays are measured by a spark-chamber magnetic spectrometer, but transformation to the K^0 c.m.s. demands a knowledge of the K^0 momentum. As in all these K^0 experiments, the K^0 beam has quite a wide spectrum of energies. However in the Princeton 3 GeV accelerator the beam is divided into a series of short bursts and the K^0 momentum measured by the time of flight, as obtained from the delay between production and detection. The $K^0 \to 2\pi^0$ decay was then determined as a fraction of the known $K^0 \to 3\pi^0$ rate by comparison of the numbers of γ-rays above and below 170 MeV. For an accurate determination of the ratio of the amplitudes

$$|\eta_{00}| = \frac{\langle \pi^0\pi^0 | H | K_2^0 \rangle}{\langle \pi^0\pi^0 | H | K_1^0 \rangle},$$

from the observed γ-rays, allowance should also be made for the contribution of the modes other than $K^0 \to \pi^0\pi^0\pi^0$ to the γ-spectrum below 165 MeV. The two experiments yield

$|\eta_{00}| = (4\cdot3^{+1\cdot1}_{-0\cdot8}) \times 10^{-3}$ Cern,

$|\eta_{00}| = (4\cdot9 \pm 0\cdot5) \times 10^{-3}$ Princeton.

We may compare this measurement of the $\pi^0 \pi^0$ decay-amplitude ratio with that for charged decays by calculating

$$|\eta_{+-}| = \frac{\pi^+\pi^- |H| K_2^0\rangle}{\pi^+\pi^- |H| K_1^0\rangle}$$

$$= \left[R \frac{\Gamma(K_2^0)_{ch}}{\Gamma(K_2^0)_{all}} \frac{\Gamma(K_2^0)_{all}}{\Gamma(K_1^0)_{all}} \frac{\Gamma(K_1^0)_{all}}{\Gamma(K_1^0)_{\pi+\pi-}} \right]^{\frac{1}{2}}$$

$$= (1\cdot95 \pm 0\cdot072) \times 10^{-3}.$$

These are then the data which any theory of the K_2^0 decay must explain.

8.6 Hypotheses concerning the $K_2^0 \to 2\pi$ decay

8.6.1 *The new-force hypothesis*

The new-force hypothesis, proposed independently by Bell and Perring, and by Bernstein, Cabibbo and Lee, postulates the existence of an external field which couples with opposite sign to particles and antiparticles. It is clear that such a field would have to be both very weak and of very long range (astronomical distances) in order to explain the phenomenon of $K_2^0 \to 2\pi$ without producing other effects which are not observed.

If the field was produced by a system itself not invariant under *CP*, for instance containing different quantities of matter and antimatter, then an *apparent CP* violation would be produced in the decays.

This interesting hypothesis has however been shown to require that the $K_2^0 \to 2\pi$ rate should depend on the K^0 energy E in the form E^{2J}, where J is the spin of the field giving rise to the effect. A further consequence of this model is that if $J = 0$ then $\eta_{+-} = \eta_{00}$ and since as we have seen this is not so we may assume that $J \geqslant 1$ so that the $K_2^0 \to 2\pi$ rate should increase at least as E^2. However the evidence from a series of different experiments, as shown in Table 5, shows clearly that the rate certainly exhibits no such dependence and indeed is perfectly consistent with being independent of E.

Table 5

$R = \dfrac{K_2^0 \to \pi^+\pi^-}{K_2^0 \to \text{all charged modes}} \times 10^3$	*Average* K_2^0 *momentum* (GeV/c)	*Reference*
$2\cdot0 \pm 0\cdot4$	$1\cdot1$	Christenson (1964)
$2\cdot08 \pm 0\cdot35$	$3\cdot15$	Galbraith (1965)
$1\cdot97 \pm 0\cdot16$	$1\cdot5$	Fitch (1967)
$2\cdot12 \pm 0\cdot18$	$10\cdot7$	de Bouard (1967)

8.6.2 *The hypothesis of CP violation in electromagnetic processes*

We have already noted that the degree of CP violation is small. Non-invariance implies

$$[H, CP] \neq 0,$$

where the Hamiltonian H consists of parts due to strong, weak and electromagnetic interactions, i.e.

$$H = H_{st} + H_\gamma + H_{wk}.$$

The smallness of the violation might then arise from violation through the small contributions to the Hamiltonian involving H_{st} or H_γ, that is through second-order terms in the interaction such as $H_\gamma H_{wk}$ or $H_{st} H_{wk}$. The proposal that the lack of invariance under CP might be due to the electromagnetic interaction has been studied particularly by Lee, who pointed out that the order of magnitude of the violation contribution by CP non-conserving electromagnetic interactions should be approximately α (or α/π) the electromagnetic coupling, ~ 7(or 2–3) $\times 10^{-3}$ as found by experiment.

In order to study this question it is necessary first to review our knowledge concerning the invariance of the strong, weak and electromagnetic interactions under C, P and T operations and their combinations. The results of this review are entered in Table 6.

Table 6

	H_{st}	H_γ	H_{wk}
CPT	YES	YES	YES
P	YES	YES	NO
C	YES	?	NO
T or CP	YES	?	?

8.7 *CPT* invariance

CPT invariance implies the equality of the masses and lifetimes of particles and their antiparticles. It may be shown that the mass difference between K^0 and $\bar{\mathrm{K}}^0$ is equal to the $\mathrm{K}_1^0 - \mathrm{K}_2^0$ mass difference Δm. (This may best be done by means of the technique of the mass matrix, described in 10.6). Thus following our earlier discussion (8.4) it is clear that the best test in respect of masses is the comparison of the K^0 and $\bar{\mathrm{K}}^0$ masses, where the difference $\Delta m/\Gamma_1$ was shown to be ~ 1 yielding $\Delta m/m \simeq 10^{-14}$. Writing for complete CPT conservation

$$\langle \mathrm{K}^0 | H_{st} + H_\gamma + H_{wk} | \mathrm{K}^0 \rangle = \langle \bar{\mathrm{K}}^0 | H_{st} + H_\gamma + H_{wk} | \bar{\mathrm{K}}^0 \rangle.$$

The above result implies invariance of H_{st} to $\sim 10^{-14}$, H_γ to $\sim 10^{-12}$ and H_{wk} to $\sim 10^{-8}$ (strangeness-conserving non-leptonic part of H_{wk}).

A comparison of particle and antiparticle lifetimes yields (see 5.3 for kaons)

$$\mu: \quad \frac{\Delta\tau}{\tau} = (0{\cdot}0 \pm 0{\cdot}1) \times 10^{-2},$$

$$\pi: \quad \frac{\Delta\tau}{\tau} = (0{\cdot}23 \pm 0{\cdot}4) \times 10^{-2},$$

$$K: \quad \frac{\Delta\tau}{\tau} = (0{\cdot}05 \pm 0{\cdot}1) \times 10^{-2},$$

again in good agreement with CPT (including the leptonic part of H_{wk}) but to a lower degree of precision.

We thus conclude that all the experimental evidence supports invariance under CPT and that the limits to which such invariance has been tested are very good.

8.8   *C*- and *P*-invariance in strong interactions

For strong interactions it is possible to test P-violation by looking for effects due to pseudo-scalar terms in low-energy nuclear processes, or in nucleon–nucleon scattering. An example of such a parity non-conserving term is a longitudinal polarization in for instance p–n scattering. A search for such polarization has been made in an experiment in which a beryllium target was bombarded by protons of 380 MeV. Neutrons emerging from the target in the line of the incident proton beam and having an energy greater than 350 MeV, were passed into a solenoid such that their spin was rotated by the magnetic field to convert any longitudinal polarization into transverse polarization. The magnitude of the resulting transverse polarization was determined by allowing the neutrons to scatter on a hydrogen target and measuring the up–down asymmetry in the neutron–proton scattering. For unpolarized neutrons no such asymmetry exists. No asymmetry was observed, and the limits of the measurement indicated that any parity non-conserving amplitude in the initial interaction must be less than about 10^{-3} of the parity conserving amplitude. Likewise no other evidence for parity non-conservation in strong processes has been found.

A more direct test has been made by seeking for parity non-conserving transitions in the decay of excited states of nuclei. For instance an excited state of ^{20}Ne with spin–parity 1^{+} may be formed by bombardment of ^{19}F by protons of appropriate energy. Such a state could decay to the ground state of ^{16}O by α-particle emission only if parity is not conserved in the transition, since ^{16}O has spin–parity 0^{+} and the α-particle spin is also zero. This transition is not observed and by this method a limit of about 10^{-6} on the relative size of the parity non-conserving amplitude has been obtained.

The best method of testing C-invariance in strong interactions is to compare the angular and energy distributions of π^{+} and π^{-}, K^{+} and K^{-}, or K^{0} and $\bar{K}^{0}$ particles in proton–antiproton annihilation into mesons. If the process is invariant under C, then the angular and energy distributions for the particle and the antiparticle should be identical. Such studies have established a limit of about 10^{-4} for the relative amplitudes for the C non-invariant and C-invariant processes.

8.9 C- and P-invariance in electromagnetic interactions

The conservation of parity in electromagnetic processes may be tested by searching for transitions between atomic states having quantum numbers such that the transitions cannot conserve parity. No such transition has been observed and an upper limit of about 10^{-3} for the relative size of the parity non-conserving amplitude in electromagnetic processes has been set by this method.

Until recently, however, no direct evidence existed concerning charge-conjugation invariance for electromagnetic processes. It was this lack of knowledge concerning C-invariance which led Lee to suggest that failure of charge-conjugation invariance in electromagnetic processes might give rise to a small C and CP non-invariant amplitude for the K_2^0 to two-pion decay through processes involving the emission and absorption of virtual gamma rays.

A number of experiments to test the hypothesis of lack of charge-conjugation invariance in electromagnetic processes have been suggested, and data from some such experiments already exists.

One such test, involving invariance under both the parity transformation and the operation of time reversal in electromagnetic processes, is the existence or otherwise of an electric-dipole moment in the neutron. We can show that the existence of such a moment implies lack of invariance under either P or T and hence, if the CPT theorem is true, also under C. The Hamiltonian for the interactions between the magnetic- and electric-dipole moments for such a particle with an electromagnetic field may be written in the form

$$H_I = \rho_m \boldsymbol{\sigma}.\mathbf{H} + \rho_e \boldsymbol{\sigma}.\mathbf{E},$$

where ρ_m and ρ_e are the magnitudes of the magnetic- and electric-dipole moments, $\boldsymbol{\sigma}$ is the spin vector and $\mathbf{H}$ and $\mathbf{E}$ are the magnetic and electric field vectors. A consideration of $\mathbf{H}$ and $\mathbf{E}$ shows that $\mathbf{H}$ is even under the parity transformation and $\mathbf{E}$ odd, while $\mathbf{H}$ is odd under time reversal and $\mathbf{E}$ is even under this operation. Also, the spin vector $\boldsymbol{\sigma}$ is even under P and odd under T, i.e. it behaves like the magnetic field in this respect. Thus the contribution to the Hamiltonian due to the magnetic-dipole moment is invariant under the parity operation *and* under time reversal, while the contribution due to the electric-dipole moment changes sign under both operations. This means that the electric-dipole moment of the neutron must be zero unless the electromagnetic interaction is not invariant under time reversal or space inversion. Since we already know that parity conservation in such interactions holds to a high degree of accuracy, the observation of such a moment would imply failure of *time reversal* invariance and thus, if CPT is to be preserved, of invariance under charge conjugation.

If we write the electric-dipole moment of the neutron in the form

$$d_n = eL_n,$$

where e is the charge on the electron and L_n is a length, then recent experiments yield values for L_n of $(-2\pm 3)\times 10^{-21}$ mm and $(2\cdot 4\pm 3\cdot 9)\times 10^{-21}$ mm, suggesting that invariance under P and thus under C is good, in contradiction to the hypothesis of Lee. Unfortunately, it is difficult to know how to calculate properly the expected value of the quantity L_n, even if Lee's hypothesis is true, so that this data cannot be taken to be conclusive.

The best studied test of C-invariance in electromagnetic processes involves the decays of the η^0 meson. This meson will be discussed in some detail in the following chapter. For the present we merely state certain of its properties as follows: mass = 549 MeV/c^2, narrow 'width' indicating a lifetime consistent with electromagnetic decay, and decay modes

$$\eta^0 \to \begin{cases} \pi^+ \pi^- \pi^0 \\ \pi^0 \pi^0 \pi^0 \\ \pi^+ \pi^- \gamma \\ \pi^0 \gamma \gamma \\ \gamma\gamma. \end{cases}$$

Even the three-pion decays probably proceed by means of electromagnetic processes (see Chapter 9) although no γ-rays are actually produced. They are found to be of similar intensity to those decays involving γ-rays.

If the decay $\eta \to \pi^+ \pi^- \pi^0$ is invariant under charge conjugation then the π^+ and π^- must behave completely alike in the final state, so that in a plot of the kind shown in Figure 58 (see also Appendix A) we expect complete symmetry between the right- and left-hand halves of the diagram. Another way of testing the symmetry is to define a parameter

$$A = \frac{N_+ - N_-}{N_+ + N_-},$$

where we have N_+ events in which the c.m.s. energy of the π^+ is greater than that of the π^- and N_- events in which the opposite is the case. For invariance under C, the parameter A must clearly equal zero.

Several experiments have tested C-violation in this way, to various degrees of accuracy. The η^0 particles may be made for instance in the reaction

$$\pi^+ d \to p\eta^0(p),$$

where the π^+ has reacted with the neutron in the deuterium and the proton has acted as a 'spectator'. The result of such an experiment, carried out by studying interactions of $0\cdot 82$ GeV/c π^+ mesons in a deuterium-filled bubble chamber, is shown in Figure 58. Each point on the diagram represents an event of the type shown above, where the η decays by $\eta \to \pi^+ \pi^- \pi^0$. Such events were identified by observation of a short-recoil proton, corresponding to the spectator recoil, followed by kinematic fitting. Each event was also examined to ensure that the bubble densities of the tracks were in agreement with the values predicted by the kinematic fit. This last process often resolved ambiguities between possible hypotheses, in particular resolving a number of cases where a fit could be achieved on the assumption that *either* of the positive tracks was the proton. In Figure 58 there are repre-

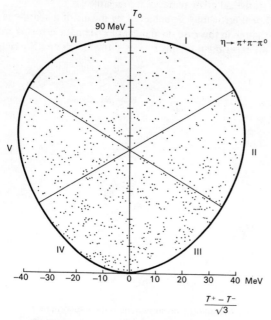

Figure 58 Dalitz plot for the 765 events of the type $\eta \to \pi^+\pi^-\pi^0$ in the experiment of Larribe *et al.* (1966). T_0, T_+ and T_- are the kinetic energies of the π^0, π^+, π^- mesons respectively in the η c.m.s.

sented 765 events. In this experiment however there were 21 000 events measured. 17 000 of these fitted the reaction

$$\pi^+ d \to p\pi^+\pi^-(p),$$

while several hundred events fitted the other processes. The number finally included in the analysis was further reduced by imposing strict criteria on the volume of the chamber used for measurement, on the minimum length for the recoil proton and on the mass accepted for the η-meson. These criteria were necessary to avoid any bias of the data due to, for instance, badly measured events which might affect the π^+ and π^- energies differently. The final value obtained for A in this experiment was

$$A = -0.048 \pm 0.036.$$

The advantage of the use of a bubble chamber for this experiment is that the possibility of biases which might produce a difference between π^+ and π^- is very small. That this is so is due to the observation of the interaction vertex itself and the essentially '4π-geometry' in which events are observed having all angles for the outgoing tracks. The disadvantage is that it is difficult to obtain a very large sample of events, so that the statistical error remains appreciable.

A much lower statistical error may be achieved by means of a spark-chamber counter experiment although in this case the danger of systematic errors is greater. The arrangement for an experiment of this type is shown in Figure 59. Here the

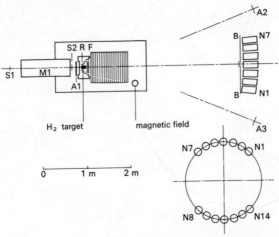

Figure 59 Schematic of the experimental arrangement in the experiment of Cnops *et al.* (1966) to study the $\eta \to \pi^+ \pi^- \pi^0$ decay. The neutrons from the production process $\pi^- p \to n\eta$ were detected in the counters N1–N14.
The counters F (with hole for beam) ensured that at least one charged particle entered the spark chambers. The counters R and A1 in anticoincidence ensure that no charged particles emerged *except* into the spark chambers. Counters S1, S2 and no count in B ensured a beam particle interaction in the hydrogen target.
M1 is a magnet to compensate the beam deflection in the main magnet surrounding the spark chambers

η-particles were produced in a liquid-hydrogen target by incident π^- mesons of 0·713 GeV/c by means of the process

$$\pi^- p \to n\eta.$$

The spark chambers detect the π^+ and π^- mesons from the η-decay to $\pi^+ \pi^- \pi^0$, while the neutrons are detected by a system of scintillation counters having their axes directed at the target. The spark chambers were triggered by a coincidence–anticoincidence arrangement which ensured that a beam particle had interacted in the target, that no charged particle had emerged other than into the spark chambers, that at least one charged particle had entered the chambers and that a neutral particle had entered the neutron counters in a specified time interval. The neutron time-of-flight was measured to yield a value for its momentum. In order to avoid any asymmetries in the result, due to asymmetries in the magnetic field or the optical system, the field was reversed during the experiment. After kinematic fitting and application of certain selection criteria, the final sample in this experiment yielded 10 665 events and the final value obtained for A was

$$A = (0 \cdot 3 \pm 1 \cdot 1) \text{ per cent.}$$

A summary of data from various experiments is given in Table 7 from which it can be seen that the data are certainly consistent with invariance of the electromagnetic interaction under C, and that any non-invariance can be no more than yields a value of $A \simeq 1$ per cent.

Table 7

A/per cent	Group
$0 \cdot 3 \pm 1 \cdot 1 \ (\pi^+ \pi^- \pi^0)$	Cnops *et al.* – spark chamber
$7 \cdot 2 \pm 2 \cdot 8 \ (\pi^+ \pi^- \pi^0)$	Baltay *et al.* – bubble chamber
$4 \cdot 1 \pm 4 \cdot 1 \ (\pi^+ \pi^- \pi^0)$	Fowler *et al.* – bubble chamber
$-0 \cdot 048 \pm 0 \cdot 036 \ (\pi^+ \pi^- \pi^0)$	Larribe *et al.* – bubble chamber
$1 \cdot 5 \pm 2 \cdot 5 \ (\pi \pi \gamma)$	Bowen *et al.* – spark chamber

In summary, it is clear that no evidence exists of C non-invariance in electromagnetic interactions and that no explanation of the CP non-invariance demonstrated in $K_2^0 \rightarrow 2\pi$ decay has yet been found.

8.10 Other selection rules for strange-particle decays: non-leptonic decays

We first consider the non-leptonic decays of the strange particles like, for instance,

$$\Lambda \rightarrow p\pi^-,$$

$$\Lambda \rightarrow n\pi^0.$$

We note that since the nucleon and pion can only combine to give total isotopic spin $I = \frac{1}{2}$ or $\frac{3}{2}$, while the i-spin of the Λ-particle is zero, therefore the i-spin cannot be conserved in this process. Gell-Mann suggested that the change in i-spin would be the minimum possible, that is that

$$|\Delta I| = \frac{1}{2}.$$

This proposal is more restrictive than the $|\Delta I_3| = \frac{1}{2}$ rule which follows from $|\Delta S| = 1$ for weak decays. We shall examine the consequences and tests of this proposed rule in the following paragraphs.

8.10.1 Λ-decay

We can write i-spin wave functions for $(p\pi^-)$ and $(n\pi^0)$ in terms of the i-spin $\frac{1}{2}$ and $\frac{3}{2}$ wave functions

$$|p\pi^-\rangle = \sqrt{\tfrac{1}{3}}|\tfrac{3}{2}\rangle - \sqrt{\tfrac{2}{3}}|\tfrac{1}{2}\rangle, \qquad\qquad 8.4$$

$$|n\pi^0\rangle = \sqrt{\tfrac{2}{3}}|\tfrac{3}{2}\rangle + \sqrt{\tfrac{1}{3}}|\tfrac{1}{2}\rangle, \qquad\qquad 8.5$$

where we have used the Clebsch–Gordan coefficients as listed in Appendix B. The relations **8.4** and **8.5** then give for the decay ratio

$$\frac{\Lambda \to p\pi^-}{\Lambda \to n\pi^0} \begin{cases} = 2:1 \text{ for } I = \tfrac{1}{2}, & \text{i.e.} \quad \Delta I = \tfrac{1}{2} \\ = 1:2 \text{ for } I = \tfrac{3}{2}, & \text{i.e.} \quad \Delta I = \tfrac{3}{2} \end{cases}.$$

As we have already seen the ratio is found to be $2:1$ supporting the $\Delta I = \frac{1}{2}$ hypothesis.

8.10.2 Σ-decay

We can apply similar arguments to Σ-decays. In this case we may, however, test the more restrictive rule that the *vector* value of I can only change by $\frac{1}{2}$, i.e.

$$\mathbf{I}_i - \mathbf{I}_f = \mathbf{I},$$

where all quantities are vectors and $\mathbf{I}$ has magnitude $\frac{1}{2}$. The $\mathbf{I}$ is sometimes referred to as belonging to a *spurion*, a fictional construct which may make it easier to think about such processes. Since the Σ-particles have i-spin 1, we expect possible final π-nucleon states having i-spin $\frac{1}{2}$ and $\frac{3}{2}$. The final state can then be represented as a combination of π, nucleon and spurion, and the amplitudes for the three charged decay channels

$$\Sigma^+ \to p\pi^0 \;(+\text{spurion}),$$

$$\Sigma^+ \to n\pi^+ \;(+\text{spurion}),$$

$$\Sigma^- \to n\pi^- \;(+\text{spurion}),$$

may be written in terms of i-spin amplitudes. We may make the combination in two steps by first treating any pair of particles and then combining the pair with the third particle. We choose first to combine the pion and nucleon since we have already treated this system (see equations **2.10** and **2.11**). Using the same notation as before we write the π–nucleon total wave function as (I, I_3) and we have

$$\left.\begin{aligned} (\tfrac{3}{2}, \tfrac{1}{2}) &= \sqrt{\tfrac{1}{3}}(\pi^+ n) + \sqrt{\tfrac{2}{3}}(\pi^0 p) \\ (\tfrac{3}{2}, -\tfrac{3}{2}) &= (n\pi^-) \\ (\tfrac{1}{2}, \tfrac{1}{2}) &= \sqrt{\tfrac{2}{3}}(\pi^+ n) - \sqrt{\tfrac{1}{3}}(\pi^0 p) \end{aligned}\right\} \qquad 8.6$$

(see Appendix B for Clebsch–Gordan coefficients).

We now combine the i-spin wave functions of **8.6** with the spurion i-spin wave function $(\frac{1}{2}, \frac{1}{2})$ to form the appropriate functions for the Σs, i.e. $(1, +1)$ and $(1, -1)$.

However we have no *a priori* knowledge of the relative strengths of the $I = \frac{3}{2}$ and $I = \frac{1}{2}$ amplitudes of equation **8.6** so that we must introduce unknown parameters x and y, being the relative amounts of the $(\frac{3}{2}, \frac{1}{2})$ and $(\frac{1}{2}, \frac{1}{2})$ states respectively. Then

$$(1, +1) = [-\tfrac{1}{2}x(\tfrac{3}{2}, \tfrac{1}{2}) + y(\tfrac{1}{2}, \tfrac{1}{2})](\tfrac{1}{2}, \tfrac{1}{2})$$

$$= [\sqrt{\tfrac{2}{3}}y(\pi^+ n) - \sqrt{\tfrac{1}{3}}y(\pi^0 p) - \tfrac{1}{2}x\sqrt{\tfrac{1}{3}}(\pi^+ n) - \tfrac{1}{2}x\sqrt{\tfrac{2}{3}}(\pi^0 p)](\tfrac{1}{2}, \tfrac{1}{2})$$

$$= [(\sqrt{\tfrac{2}{3}}y - \sqrt{\tfrac{1}{12}}x)(\pi^+ n) - (\sqrt{\tfrac{1}{3}}y + \sqrt{\tfrac{1}{6}}x)(\pi^0 p)](\tfrac{1}{2}, \tfrac{1}{2}).$$

.Similarly

$$(1, -1) = -x\sqrt{\tfrac{3}{4}}(\pi^- n)(\tfrac{1}{2}, \tfrac{1}{2}).$$

Then if we forget the spurion, which was introduced only to simplify the manipulation of $\Delta I = \frac{1}{2}$, we can write the relative (complex) amplitudes for the decay modes of the Σs

$$\mathbf{T}_+(\Sigma^+ \to n\pi^+) = \sqrt{\tfrac{2}{3}}y - \sqrt{\tfrac{1}{12}}x, \qquad \textbf{8.7}$$

$$\mathbf{T}_0(\Sigma^+ \to p\pi^0) = -\sqrt{\tfrac{1}{3}}y - \sqrt{\tfrac{1}{6}}x, \qquad \textbf{8.8}$$

$$\mathbf{T}_-(\Sigma^- \to n\pi^-) = -\sqrt{\tfrac{3}{4}}x. \qquad \textbf{8.9}$$

Eliminating y from the first two equations we then have a triangular relationship

$$\mathbf{T}_+ + \sqrt{2}\mathbf{T}_0 = \mathbf{T}_-. \qquad \textbf{8.10}$$

In addition we have seen that the lifetimes for the three decays **8.7**, **8.8**, **8.9** are practically equal, so that

$$|\mathbf{T}_+|^2 = |\mathbf{T}_0|^2 = |\mathbf{T}_-|^2,$$

since the phase-space factors in all the decays are also very similar. This equality of the moduli combined with the relation **8.10** implies that the **T**-vectors form a right-angled triangle.

Further information concerning the angles between the amplitude vectors may be obtained from the decay angular-distributions of polarized hyperons. The treatment of this topic falls beyond the scope of the present book, but the data available show that in fact the vectors fail to fit the triangular relationship by an amount beyond the experimental errors, suggesting that there may be present some $|\Delta I| = \frac{3}{2}$ amplitude.

8.10.3   K-*meson decays*

We first consider the decays of a kaon into two pions, such as

$$\mathrm{K}^0_1 \to \begin{cases} \pi^+ \pi^- \\ \pi^0 \pi^0 . \end{cases}$$

The $|\Delta I| = \frac{1}{2}$ rule allows final states having $I = 0$ or 1. A $|\Delta I| = \frac{3}{2}$ transition would also allow access to the $I = 2$ state. In addition, since the mesons involved are all spin-zero particles, the two pions must be in an s-state. The pions are bosons, so the overall wave function must be symmetric. Thus the i-spin wave function must also be symmetric so that I must be even, and if $|\Delta I| = \frac{1}{2}$ then I must be zero.

Referring once more to the Clebsch–Gordan coefficients, we write the $I = 0$ wave function for two pions as

$$\frac{1}{\sqrt{3}} \left(\pi_1^+ \pi_2^- - \pi_1^0 \pi_2^0 + \pi_1^- \pi_2^+ \right),$$

yielding the ratio

$$\frac{K_1^0 \to \pi^+ \pi^-}{K_1^0 \to \text{All}} = \frac{2}{3},$$

in very good agreement with experiment.

However we immediately see that for

$$K^\pm \to \pi^\pm \pi^0,$$

since $I_3 = 1$ for the final state, then, using our previous argument, I can only equal 2 so that $|\Delta I| = \frac{3}{2}$. Nevertheless we note that the decay rate for the neutral K-decay is about 700 times faster (see Appendix B) than that for the charged kaon, so that although not inviolable the $|\Delta I| = \frac{1}{2}$ rule is fairly strong. The presence of a $|\Delta I| = \frac{3}{2}$ amplitude would also affect the $K^0 \to 2\pi$ decay ratios, but since the amplitude is small this effect is such that it is within the errors of the measurements.

8.11 Leptonic decays of strange particles

As elsewhere, the discussion here is solely concerned with the conservation laws effective in these decays.

In Chapter 5 we saw that both the hyperons and the K-mesons had decay modes involving leptons, such as for instance

$$\Lambda \to p + e^- + \nu, \qquad \qquad \textbf{8.11}$$
$$K^0 \to \pi^- + e^+ + \nu. \qquad \qquad \textbf{8.12}$$

The branching ratios for the leptonic modes for the hyperons are small, so that they are difficult to study. Nevertheless, data now available allow some interesting conclusions regarding these decays.

Possibly the most productive way to study these modes has been by producing low-energy hyperons in very large numbers, in a liquid-hydrogen bubble chamber. In a recent experiment of this kind carried out by a group from Heidelberg, using the CERN accelerator, $10^7 K^-$ mesons were stopped in a hydrogen chamber. Candidates for leptonic decays of hyperons were selected by a crude measurement, using fast rough digitizers connected on-line to a computer. Events passing this rough filter were then studied in detail with measurements of high precision. Heavy-liquid bubble chambers have also been used in studies of this kind.

As for the non-leptonic modes, we have $|\Delta S| = 1$, so that we again have from

$$Q = I_3 + \frac{B + S}{2}$$

that $\Delta Q = \frac{1}{2}\Delta S + \Delta I_3$.

Note that here the changes in Q, S and I_3 refer to the *strongly interacting particles* for which the original relationship holds and since we also have additional charged particles present, ΔQ need no longer be zero. Since ΔQ and ΔS can each equal ± 1, two possibilities apparently exist

$$\Delta Q = \Delta S \quad \text{i.e.} \quad \frac{\Delta Q}{\Delta S} = 1,$$

$$\Delta Q = -\Delta S \quad \text{i.e.} \quad \frac{\Delta Q}{\Delta S} = -1.$$

For $\quad \Delta Q = \Delta S, \quad |\Delta I_3| = \frac{1}{2} \quad$ and $\quad |\Delta I| \geqslant \frac{1}{2}$
while for $\quad \Delta Q = -\Delta S, |\Delta I_3| = \frac{3}{2} \quad$ and $\quad |\Delta I| \geqslant \frac{3}{2}$.

Note that although $|\Delta I| = \frac{1}{2}$ implies $\Delta Q = \Delta S$ the converse is not true. The processes **8.11** and **8.12** clearly require i-spin changes for the strongly interacting particles on each side of the equation.

We may consider some possible leptonic decay processes the presence or apparent absence of which leads us to the conclusion that the '$\Delta S = \Delta Q$ rule' is true.

Allowed by the rule are processes:

$$\Lambda \rightarrow pe^-\bar{\nu} \qquad \Delta S = \Delta Q = +1$$
$$\Sigma^- \rightarrow ne^+\nu \qquad \Delta S = \Delta Q = +1$$
$$\Xi^- \rightarrow \Lambda e^-\bar{\nu} \qquad \Delta S = \Delta Q = +1$$
$$K^0 \rightarrow \pi^-e^+\nu \qquad \Delta S = \Delta Q = -1$$
$$\quad \pi^-\mu^+\nu$$
$$\bar{K}^0 \rightarrow \pi^+\mu^-\bar{\nu} \qquad \Delta S = \Delta Q = +1$$
$$\quad \pi^+e^-\bar{\nu}$$
$$K^+ \rightarrow \pi^+\pi^-e^+\nu \qquad \Delta S = \Delta Q = -1.$$

Not allowed are:

$$\Sigma^+ \rightarrow ne^+\nu \qquad \Delta S = -\Delta Q = +1$$
$$K^0 \rightarrow \pi^+e^-\bar{\nu} \qquad \Delta S = -\Delta Q = -1$$
$$\quad \pi^+\mu^-\bar{\nu}$$
$$\bar{K}^0 \rightarrow \pi^-e^+\nu \qquad \Delta S = -\Delta Q = +1$$
$$\quad \pi^-\mu^+\nu$$
$$K^+ \rightarrow \pi^+\pi^+e^-\bar{\nu} \qquad \Delta S = -\Delta Q = -1.$$

Although a possible example of $\Sigma^+ \rightarrow ne^+\nu$ has been reported, it is quite clear that this process is at least severely inhibited relative to the corresponding $\Delta Q = \Delta S$ decay. Also for the K^+_{e4} decays ($K \rightarrow \pi\pi e\nu$), although this is a rare decay mode, several hundred examples have now been studied in heavy-liquid bubble chambers with no instance of the unallowed $K^+ \to \pi^+ \pi^+ e^- \bar{\nu}$ process having been seen.

For the K^0 and $\bar{K}^0$ particles the leptonic decay modes have large branching ratios for the long-lived fraction but, due to the complex nature of the K^0–$\bar{K}^0$ system, the testing of the $\Delta Q = \Delta S$ rule for these particles is slightly less direct than for the charged Σs.

If the rule is true then we note that only $\bar{K}^0$ can decay by the electron mode, so that the ratio of positron to electron decays should follow the ratio of K^0 to $\bar{K}^0$ in a K^0–$\bar{K}^0$ 'beam'.

If we refer to section 8.4, equations **8.2** and **8.3** give expressions for the numbers of K^0 and $\bar{K}^0$ as a function of time. If we start off with a pure beam, and if $\Delta S = \Delta Q$, the ratio of these expressions should give the ratio of positron to electron decays. A departure from the $\Delta S = \Delta Q$ rule would allow K^0 to decay into electrons and $\bar{K}^0$ to decay into positrons, producing an effect which we parameterize by introducing a factor x, so that (we assume CP conservation)

$$\frac{N_{e^+}(t)}{N_{e^-}(t)} = \frac{[(1+x)e^{-\Gamma_1 t} + (1-x)e^{-\Gamma_2 t} + 2\cos\Delta m\, e^{-\frac{1}{2}(\Gamma_1+\Gamma_2)t}]}{[(1+x)e^{-\Gamma_1 t} + (1-x)e^{-\Gamma_2 t} - 2\cos\Delta m\, e^{-\frac{1}{2}(\Gamma_1+\Gamma_2)t}]},$$

where $x = 0$ gives the expected time variation if the rule is true.

An alternative test is to study the time variation of the sum of positron and electron decays as a function of time

$$N_{e^+}(t) + N_{e^-}(t) \propto [1+x)e^{-\Gamma_1 t} + (1-x)e^{-\Gamma_2 t}].$$

The test for K^0 mesons is thus not a simple one of search for examples of forbidden processes, but involves study of the decays as a function of time. The experiments are difficult since one does not have a 'beam' of K^0, in the normal sense of the word, but quite a limited number of decays normally observed in a heavy-liquid bubble chamber. For these reasons, and because of purely experimental difficulties, it has not been possible to achieve an unambiguous result. The results from various experiments have generally large errors and do not even agree within these large error limits. The question of the correctness of the $\Delta S = \Delta Q$ rule for the K^0 leptonic decays is thus at present unanswered.

Chapter 9
Strongly decaying resonances

9.1 Introduction

In the course of the foregoing chapters we have studied unstable particles which decayed via the *weak* and *electromagnetic* interactions. In particular, in dealing with strangeness, it was shown that the strange-particle decays had lifetimes generally in the region of 10^{-8}–10^{-10} s, characteristic of the weak interaction. We also have examples of electromagnetic decay processes involving γ-rays, such as

$$\pi^0 \to \gamma\gamma,$$

$$\Sigma^0 \to \Lambda\gamma,$$

for which the lifetimes are $\sim 10^{-15}$–10^{-16} s. It is then natural to ask whether there are particles which decay via the *strong* interaction. Since we know from consideration of the cross-section magnitudes that the strong interaction is some 10^{14} times stronger than the weak interaction, we might expect lifetimes for strong decay processes of about 10^{-23} s.

If particles decaying via strong interactions exist they may therefore be expected to have lifetimes such that, even if travelling near the speed of light, the distance they travel before decay is so short as to make it quite impossible to distinguish production and decay points. Thus in seeking such particles we cannot for instance expect to see a track in a bubble chamber but must rely on indirect evidence. Applying the uncertainty relationship in the form

$$\Delta E\, \Delta t \simeq \hbar \qquad\qquad 9.1$$

to possible particles with lifetimes (i.e. Δt) $\tau \simeq 10^{-23}$ s, it is clear that the width or uncertainty in mass may be appreciable. For $\tau \simeq 10^{-23}$ s, ΔE is about 100 MeV. This means that for such a particle we might expect to find masses differing by energies of this order.

9.2 The Δ^{++} (1238): formation experiments

The discussion of isotopic spin in Chapter 2 revealed that there was a particularly strong interaction between π-mesons and nucleons in the state of total i-spin $I = \frac{3}{2}$ when the incident-pion kinetic energy was about 180 MeV. Experimentally, this result is deduced from a substantial peak in the $\pi^+ p$ elastic-scattering cross-section (as well as the values of the $\pi^- p$ elastic and charge-exchange cross-sections). We may interpret this strong interaction at a particular incident energy as being equivalent to the formation of a well-defined particle, resonance, or nucleon isobar, of very short life, according to a diagram of the form:

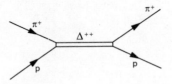

Figure 60

The resonance then acts as a short-lived particle having $I = \frac{3}{2}$, $I_3 = +\frac{3}{2}$ and charge $Q = +2$. The mass is the total centre-of-mass energy at which the resonance peak occurs. For $T_\pi = 180$ MeV we find $P_\pi = 288$ MeV/c so that

$$E^* = \sqrt{(E^2 - p^2)} = \{(0\cdot 18 + 0\cdot 14 + 0\cdot 94)^2 - (0\cdot 288)^2\}^{\frac{1}{2}} \text{ GeV}$$
$$= 1\cdot 23 \text{ GeV} = M.$$

Looking again at Figure 31 on page 70 we see that the width of the resonance peak is about 120 MeV, so that according to the uncertainty principle argument the lifetime must indeed be about 10^{-23} s, characteristic of strong decay.

Such a method of studying resonance particles is known as a 'formation' experiment. For such an experiment the cross-section for a given process, like πp elastic scattering, is studied as a function of incoming particle momentum. In the simplest case the formation of an intermediate-state resonance particle is indicated by a peak in the total cross-section.

This kind of process is sometimes referred to as the appearance of a resonance in the 's-channel' where the symbol s is used for the total energy in the c.m.s.

9.3 Resonance spin and parity: the Δ (1238)

It is not the intention to develop here the full theory of scattering in terms of partial waves, which may be found in any standard text on quantum mechanics. Rather we present a brief treatment designed to bring out the principal features.

In a study of a resonance in the s-channel, such as the Δ^{++} in $\pi^+ p$ scattering, the differential-scattering cross-section will be a function of the angular momentum of the intermediate state, i.e. the resonance spin.

Since the pion spin is zero, the angular momenta to be combined in the initial state are the proton spin and the orbital angular momentum l ($\hbar = 1$ throughout). In the initial state the proton spin can be 'up' or 'down' so that the angular momentum j of the state is $j = l \pm \frac{1}{2}$. We note that the proton spin may be 'flipped' in the interaction. In this case, since j is conserved, a change $\Delta s = \pm 1$ in the z-component of the proton spin implies $\Delta l_z = \pm 1$ since $\Delta j_z = 0$. We make our

z-direction along the line of flight of the pion and proton in the c.m.s. so that $l_z = 0$ in the initial state, and for spin flip l_z (final) $= \pm 1$.

The scattered amplitude $f(\theta, \phi)$ can then be written as a sum over the associated Legendre polynomial $Y_{l,m}$ defined in the usual way as

$$Y_{l,m}(\theta, \phi) = C_{l,m} P_l^m (\cos \theta) e^{im\phi},$$

where θ is the usual polar scattering angle, ϕ is the azimuthal angle and C is a function of m and l. For a polarized target, a situation in which more protons are aligned in one direction than in the other may be attained by means of a magnetic field acting on a suitable molecule at very low temperature. For such a target the ϕ-distribution may be anisotropic, but otherwise the angular distribution will always be isotropic in the azimuthal angle. Thus, integrating over ϕ, we write

$$f(\theta) \propto \sum_l C(l, m, E^*) P_l^m (\cos \theta),$$

where E^* is the total centre-of-mass energy and the Cs are weighting factors. The spin-flip terms have $m = \pm 1$, while the no-flip terms have $m = 0$. Thus we see that spin-flip and no-spin-flip interactions give different θ-distributions.

If $l = 0$ (s-wave scattering), so that the intermediate state has spin $\frac{1}{2}$, then only $Y_{0,0} = 1/\sqrt{(4\pi)}$ can enter, so that the angular distribution is isotropic.

If $l = 1$, then $j = \frac{1}{2}$ or $\frac{3}{2}$, i.e. we can have $p_{\frac{1}{2}}$ or $p_{\frac{3}{2}}$ states. Each of these states can include spin-flip and no-flip terms. Let us write the orbital angular-momentum wave function for the pion–nucleon combination as $\psi(l, m)$ and the nucleon-spin wave function as $\chi(S, S_z)$. The $\psi(l, m)$ are simply proportional to our $Y_{l,m}$. We arbitrarily choose the initial proton spin 'up', $S_z = +\frac{1}{2}$ since whichever choice we may make is irrelevant, the flip or no-flip being the significant factor. For this choice the initial state may be written as proportional to

$$\psi(l, 0) \chi(\tfrac{1}{2}, +\tfrac{1}{2}).$$

The final state with $j = \frac{3}{2}$, $j_z = \frac{1}{2}$ is then

$$\phi(\tfrac{3}{2}, \tfrac{1}{2}) = \sqrt{\tfrac{1}{3}}\psi(1, 1) \; \chi(\tfrac{1}{2}, -\tfrac{1}{2}) + \sqrt{\tfrac{2}{3}}\psi(1, 0) \; \chi(\tfrac{1}{2}, \tfrac{1}{2}),$$

where we have used the Clebsch–Gordan coefficients for combination of angular momenta 1 and $\frac{1}{2}$. Also

$$\phi(\tfrac{1}{2}, \tfrac{1}{2}) = \sqrt{\tfrac{2}{3}}\psi(1, 1) \; \chi(\tfrac{1}{2}, -\tfrac{1}{2}) - \sqrt{\tfrac{1}{3}}\psi(1, 0) \; \chi(\tfrac{1}{2}, \tfrac{1}{2}).$$

Thus both these states involve $Y_{1,1}$ and $Y_{1,0}$ terms, but in different ways. A pure $p_{\frac{1}{2}}$ state would give an amplitude proportional to

$$\alpha \sqrt{\frac{3}{4\pi}} \cos \theta - \beta \sqrt{2} \sqrt{\frac{3}{8\pi}} \sin \theta \, e^{i\phi},$$

where we have written α and β for $\chi(\tfrac{1}{2}, \tfrac{1}{2})$ and $\chi(\tfrac{1}{2}, -\tfrac{1}{2})$ respectively. The pure $p_{\frac{3}{2}}$ state has amplitude proportional to

$$\alpha \sqrt{2} \sqrt{\frac{3}{4\pi}} \cos \theta + \beta \sqrt{\frac{3}{8\pi}} \sin \theta \, e^{i\phi}.$$

The spin-flip and no-flip amplitudes are orthonormal, and do not interfere, so that we add the squares of the amplitudes to get for the $p_{\frac{1}{2}}$ state

$\cos^2\theta + \sin^2\theta \rightarrow$ isotropy,

and for the $p_{\frac{3}{2}}$ state

$2\cos^2\theta + \frac{1}{2}\sin^2\theta \propto 1 + 3\cos^2\theta,$

where we have dropped the nucleon-spin wave functions.

Our conclusion is that for a pure spin state decaying via the strong interaction we expect an isotropic angular distribution for a spin-$\frac{1}{2}$ state, and a distribution proportional to $1 + 3\cos^2\theta$ for a spin-$\frac{3}{2}$ state. In fact, at the Δ (1238) resonance in π^+p elastic scattering, the differential cross-section has a form very close to $1 + 3\cos^2\theta$. Some deviation from such a distribution can arise from the existence of a background of events which have not passed through the intermediate resonance state.

Since the resonance state is $p_{\frac{3}{2}}$ the parity is

$$P(\pi)\, P(p)\, (-1)^l = +1.$$

In some cases a useful clue to the spin of a resonance state may be obtained from the total cross-section. The partial-wave analysis leads to an expression for the total elastic cross-section

$$\sigma_T = \frac{2\pi}{k^2} \sum_l (2j+1)\sin^2\delta_l,$$

where $\hbar k$ is the momentum of the incoming particle, $j\hbar$ is the total angular momentum and δ_l is the 'phase shift' for the wave with orbital angular momentum $l\hbar$. Suppose that the interaction takes place predominantly in a single state of well-defined angular momentum $l\hbar$, as in the case of an s-channel resonance. In that case there is only one important term in the summation and

$$\sigma_T = \frac{2\pi}{k^2} (2j+1)\sin^2\delta_l.$$

The maximum value of σ_T occurs for $\delta_l = \frac{1}{2}\pi$. Thus the maximum value of the total cross-section for resonance in a given partial wave is

$$\frac{2\pi}{k^2} (2j+1),$$

and for $j = \frac{3}{2}$ we have an upper limit of $8\pi/k^2$ for the elastic scattering via an intermediate state of spin $\frac{3}{2}$. The limit $8\pi/k^2$ is drawn on Figure 61 and it is seen that the cross-section does indeed reach just this limit at the resonance energy.

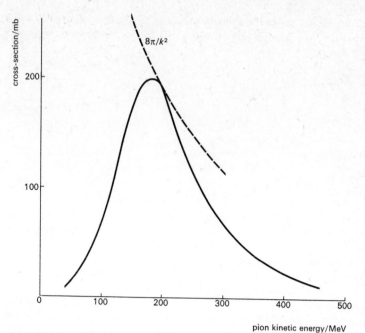

Figure 61 The π^+p scattering cross-section showing the maximum cross-section $8\pi/k^2$ for scattering in the $j = \frac{3}{2}$ state

9.4 'Production' experiments

The Δ^{++} (1238) was the first resonance found in a formation experiment, but its status as a strongly decaying very short-lived particle was only recognized when it was also found in 'production' in studies of the reaction pp → pnπ^+.

Since there are three particles in the final state, each one may have a range of momenta within the limits of energy and momentum conservation. As we have seen in other applications the transition probability can be written as

$$T = \frac{2\pi}{\hbar} |M|^2 \frac{dN}{dE}.$$

If $|M|^2$ is not a function of the individual particle momenta, then the momentum distribution of pion, proton and neutron will each range throughout the values allowed by energy and momentum conservation, with probabilities set by the density-of-states or 'phase-space' factor dN/dE. The dependence of dN/dE on the momenta is derived, for some simple cases, in Appendix A. If, on the other hand, there is a strong interaction between two of the particles for certain momenta, this would be reflected by a strong dependence of M on these quantities.

In particular, suppose that the π^+ and the proton form a Δ (1238) which lives for $\sim 10^{-23}$ s before decaying. In such a case the primary reaction is in fact a two-body process

$$pp \rightarrow n\Delta^{++} \rightarrow np\pi^+,$$

as illustrated in Figure 62.

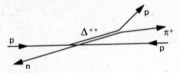

Figure 62

In the c.m.s., however, the momenta of the product particles in a two-body process are uniquely constrained, so that we should expect a unique momentum for the neutron instead of the phase-space distribution. In practice, since the Δ (1238) has a substantial width or spread in mass, the neutron momentum might be expected to have a corresponding spread. The distribution of the neutron momentum in this reaction for protons of 2·8 GeV/c is shown in Figure 63. The disagreement with the 'phase-space' distribution is obvious and it is qualitatively equally clear that the distribution might be fitted by a phase-space contribution *plus* a peak centred at a momentum of 1·62 GeV/c. The reader can check that this momentum corresponds to recoil of the neutron against a particle of mass 1236 MeV/c^2.

An alternative way of studying resonances in production experiments is by an examination of the distribution of the 'effective mass' of a group of particles. Using the general relativistic relation between total energy E, momentum p and mass m $E^2 - p^2 = m^2$, we have for the effective mass of a group of i particles

$$m_{1,\ldots,i}^2 = \left(\sum_{n=1}^{i} E_n \right)^2 - \left(\sum_{n=1}^{i} \mathbf{p}_n \right)^2.$$

We may calculate the expected distribution of m^2 if the matrix element does not depend on these quantities, that is if the transition probability depends only on the density-of-states or phase space factor. For the three-particle case, such phase-space calculations are presented in Appendix A. The phase-space distribution can be calculated analytically for up to four particles in the final state, but for larger numbers the integrals cannot be evaluated directly. However a recursion relation may be used to relate the phase space for n bodies to that for $n-1$ bodies or a 'Monte Carlo' calculation may be carried out in which artificial events are generated in a random way, but with energy and momentum conservation imposed. In all cases the phase-space distribution is smooth without any narrow peaks. If however a short-lived resonance exists in the mass combination $m_{1,\ldots,i}$ then the distribution will show a peak at this mass.

For an unstable particle decaying according to an exponential law, the shape of the distribution in mass takes what is known as a Breit–Wigner peak. For a state

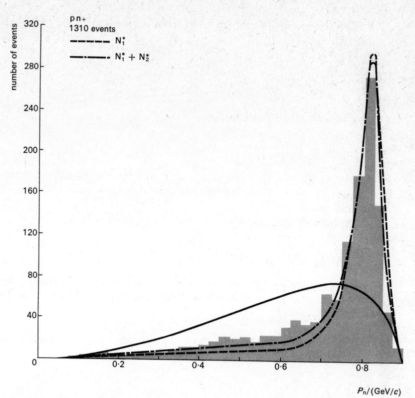

Figure 63　C.M.S. distribution of the neutrons from the reaction pp → pn π⁺ at 2·81 GeV/c. The continuous curve is that expected if the distribution is governed by phase space only. The dotted curve is that expected for Δ (1236) ($I = \frac{3}{2}$) production only. The dashed curve is that including both this resonance and an $I = \frac{1}{2}$N* (Fickinger *et al.*, 1962)

decaying exponentially with a mean life τ, we may write a wave function where the usual time-dependent part is multiplied by an exponential factor

$$\psi(t) = e^{-t/2\tau}\, e^{-iE_\mathrm{R}t/\hbar},$$

where E_R is the resonance energy. What we need is the wave function in terms of energy so we take the Fourier transform and get

$$\phi(E) = \int_0^\infty \psi(t)\, e^{iEt/\hbar}\, dt$$

$$= \int_0^\infty e^{-\frac{1}{2}t - it(E_\mathrm{R} - E)/\hbar}\, dt$$

$$= \frac{-i\hbar}{(E_\mathrm{R} - E) - \frac{1}{2}i\hbar\tau^{-1}}.$$

Writing the uncertainty relation **9.1** in terms of the width of the peak Γ and the mean life as

$$\Gamma\tau \simeq \hbar$$

we have $\quad \phi(E) = \dfrac{i\hbar}{(E_R - E) - \frac{1}{2}i\Gamma}.$ $\qquad\qquad$ **9.2**

Thus we may expect a distribution

$$|\phi(E)|^2 \propto \frac{1}{(E_R - E)^2 + \frac{1}{4}\Gamma^2}.$$

At $E = E_R$ we have the maximum $\propto \Gamma^{-2}$ while at $E = E_R \pm \frac{1}{2}\Gamma$ we get a value half that at maximum, in line with the usual definition of Γ as the full width at half-maximum.

The Breit–Wigner amplitude of **9.2** when expressed in the form

$$\phi(E) \propto \frac{1}{(E_R - E)^2 + \frac{1}{4}\Gamma^2}\left[(E_R - E) + \frac{i\Gamma}{2}\right]$$

can be conveniently represented as a function of energy on the Argand diagram. In the Argand diagram the equation of a circle of centre $(0, \frac{1}{2}i)$ and radius $\frac{1}{2}$ is

$$x^2 + (y - \tfrac{1}{2})^2 = \tfrac{1}{4},$$

and we may see that the Breit–Wigner amplitude sweeps out such a circle by writing it in the form

$$\phi(E) = \frac{2}{\Gamma}(x + iy),$$

where $\quad x = \dfrac{(E_R - E)\frac{1}{2}\Gamma}{(E_R - E)^2 + \frac{1}{4}\Gamma^2} \quad$ and $\quad y = \dfrac{\frac{1}{2}\Gamma\frac{1}{2}\Gamma}{(E_R - E)^2 + \frac{1}{4}\Gamma^2}$

(see Figure 64).

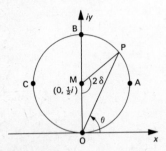

Figure 64

At A, $E = E_R - \frac{1}{2}\Gamma$, at B, $E = E_R$ and at C, $E = E_R + \frac{1}{2}\Gamma$. Thus as E increases through the resonance the amplitude vector OP sweeps out a counterclockwise circle.

The passage of the vector counterclockwise through the purely imaginary value at B is a necessary condition for resonance, although recent studies have shown that causes other than resonances may also result in this kind of behaviour. The phase shift δ may be shown to be half the angle OMP so that at resonance $\delta = \frac{1}{2}\pi$.

The amplitude passes rapidly through the upper half of the circle compared with the regions O to A and C to O. The velocity of $d\phi/dE$ is a maximum at B, and this condition is often useful in identifying the resonance energy in a case where the resonance is superimposed on a non-resonant background. In such an event the amplitude will no longer describe a simple circle but the circle will be super-imposed on a background variation with results as shown in Figure 65. A further

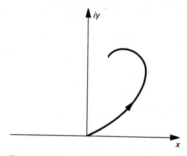

Figure 65

modification to the simple theory outlined above is necessary if the resonance is not purely 'elastic'. Such a situation may arise if a resonance has several decay channels. Suppose for instance that in a formation experiment of pions on protons one can produce a Δ resonance which can decay both by the 'elastic' channel, but also into a nucleon and two pions or a Λ^0 plus K^0

$$\pi^- p \longrightarrow \Delta \begin{array}{l} \longrightarrow \pi^- p \quad \text{(a)} \\ \longrightarrow \pi^- p\pi^0 \quad \text{(b)} \\ \searrow \Lambda^0 K^0 \quad \text{(c)} \end{array} \qquad \textbf{9.3}$$

The full width Γ for the process **9.3** is then the sum of the three partial widths Γ (a), Γ (b) and Γ (c). In such a case, the elastic amplitude sweeps out a path *within* the 'unitary circle' rather than along it. If the resonant amplitude is less than the radius of the unitary circle then the phase shift at resonance is 0° rather than $\frac{1}{2}\pi$.

9.5   The Dalitz plot

A particularly useful technique in the study of resonances in production experi-ments involving three particles in the final state is the Dalitz plot, discussed in

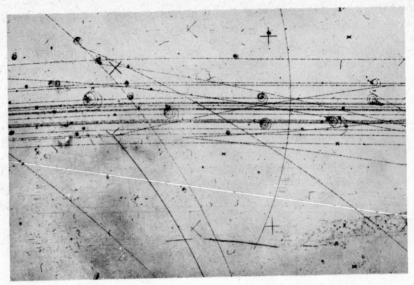

Figure 66 A reaction of the type $K^+p \to K^0\pi^+p$, using 10 GeV/c kaons in a hydrogen bubble chamber. The second beam track from the bottom is seen to interact yielding two charged tracks and a V^0 in the forward direction

detail in Appendix A and already introduced in the discussion of the τ-decay (see 7.4) to which it was first applied.

We may illustrate the application of this technique particularly well by studying, for instance, the process

$$K^+p \to K^0\pi^+p. \qquad\qquad 9.4$$

This process has been extensively studied in bubble chambers over a wide range of incident energies. A separated beam of K^+ mesons is produced, by means of the techniques described in Chapter 1, and directed into the chamber. Reactions of the kind **9.4** have the form shown in Figure 66 and may be identified by measurement, reconstruction in space and kinematic analysis. A Dalitz plot can then be constructed by plotting the kinetic energies of any two particles one against the other or, as has become more usual, by plotting the squares of the effective masses of two pairs of particles.

In Figure 67 is shown the Dalitz plot for this process with $m^2(K^0\pi^+)$ plotted against $m^2(p\pi^+)$ for an incident kaon momentum of 10 GeV/c.

Figure 67 The Dalitz plot and mass distribution for the reaction $K^+p \to K^0\pi^+p$ at 10 GeV/c. The Δ (1238) and K^* (890) and K^* (1400) are clearly visible

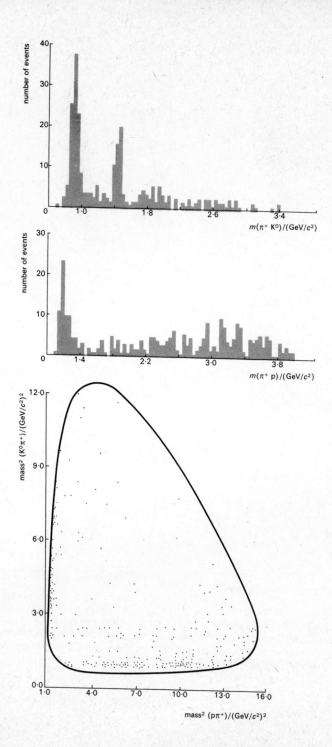

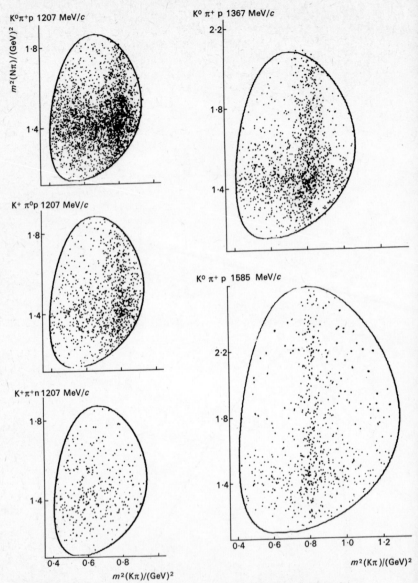

Figure 68 Dalitz plots for the process $K^+p \to K^0p\pi^+$ at various momenta showing interference between K^{*+} (890) and Δ^{++} (1236). A plot for the process $K^+p \to K^+n\pi^+$ is shown for comparison. Note that in this case there is no K^* band and the Δ is very weak (Bland et al., 1969)

From the evidence already presented for the existence of the Δ^{++} (1238) we should expect a concentration in a band centred at $(1\cdot236)^2$ on the $p\pi^+$ axis and indeed such a band is observed. We might also expect some uniform phase-space background but at this energy the background is seen to be quite small.

Our discussion to date however has not suggested the existence of the excited meson states, two of which are clearly visible in Figure 67 as bands centred at $(0\cdot890)^2$ and $(1\cdot420)^2$ on the $K^0\pi^+$ axis. These are known as the K^{*+} (890) and K^{*+} (1420) resonances. They have strangeness $+1$, baryon number zero, and decay into $K^0\pi^+$, into $K^+\pi^0$ and in the case of the K^* (1420) also into the $K\pi\pi$ final state.

We see that the resonance bands on the plot may overlap. In such an event we have two different amplitudes (for instance those for K^* (890) and Δ (1236)) leading to the same final state. This gives the possibility of interference, yielding an excess or deficit of particles in the overlap region over what might be expected by a simple addition of bands. A good example of this effect is shown in Figure 68 where results from the same process are presented but for a lower incident momentum. The accessible kinematic region is of course different, the K^* (1420) is not accessible, the K^* (890) and Δ (1236) bands overlap and the overlap region shows constructive interference.

The Dalitz plot as such may only be used for a three-particle final state. However, a modification of the plot is often useful for a larger number of final particles. For instance consider the reaction

$$K^+p \rightarrow p\pi^+K^+\pi^-.$$

We expect to find resonances in the $p\pi^+$ and $K^+\pi^-$ systems, so that it is natural to examine the structures in this process by plotting $m(p\pi^+)$ against $m(K^+\pi^-)$. In this case the boundary of the kinematically allowed region is a triangle as shown in Figure 69. The Δ and K^* (890) bands show clearly, and there is a marked interference in the overlap of the bands. It should be remembered that for this type of plot a phase-space background will not yield a uniform distribution over the allowed region. Nevertheless, the marked bands cannot be due to phase-space variations since calculation shows that the phase-space density varies smoothly over the plot, and indeed only changes rapidly near the edges of the allowed region on which it falls to zero.

9.6 Resonances with $B = 1$ and $S = 0$

The Δ^{++} (1236), which we have discussed as an example in the preceding sections, is perhaps the most readily produced of all the strangeness-zero baryonic resonances. We have seen that it has isotopic spin $I = \frac{3}{2}$, $I_3 = \frac{3}{2}$. This means that it is an i-spin quadruplet ($N = 2I+1$) and that we might expect to find charge states $+, 0, -$ as well as $+ +$. These states are found in formation experiments, and they are also seen in production.

Many other Δ or N* resonances have been found, nearly all in formation experiments with pions on nucleons. Most of these resonances are not very evident as actual peaks in cross-sections, but may appear, for instance, as enhancements in particular terms in, say, the elastic-scattering angular distribution, when a resonance of a particular spin is evidenced as an enhancement in the appropriate partial wave. In practice the study of many of these resonances has been by means of a phase-shift analysis in which the angular distributions are analysed in terms of the amplitudes of the partial waves. The complex amplitudes may then exhibit resonances according to the criteria we have already discussed.

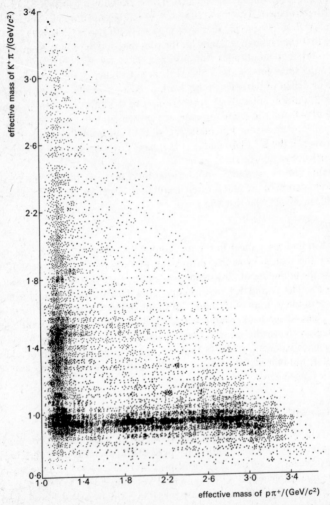

Figure 69 An example of a triangle plot for the process $K^+p \rightarrow K^+p\pi^+\pi^-$ for 10 GeV/c kaons. The plot, representing 8331 events, shows strong horizontal and vertical bands due to the K^* (890) (decaying to $K^+\pi^-$) and Δ^{++} (1236) resonances respectively

The Δ (1236) is, in many experiments, more evident than the other N*s not only because it is often strongly produced but also because of its low Q-value. Thus it occurs near the lower end of phase space. In formation experiments one also has the advantage, for $I = \frac{3}{2}$ resonances, of having available a pure $I = \frac{3}{2}$ initial state. The study of the higher-mass resonances is always further complicated by the presence of the 'tails' of those of lower mass.

A nomenclature which has become commonly accepted is that $I = \frac{1}{2}$, $S = 0$ resonances are called N*s, and $I = \frac{3}{2}$, $S = 0$ resonances are called Δs. So far the evidence for the existence of resonances with $I > \frac{3}{2}$ is very meagre. We shall return to the classification of these particles, and the proposed reason for the absence of higher i-spins, in the following chapter. The known resonances with $B = 1$ and $S = 0$ are listed in Table 8. It is very likely that continuing experiments will show this list to be incomplete.

Table 8 Baryon resonances with strangeness zero

	Resonance	J^P	Mass/ (MeV/c^2)	Width (Γ)/ (MeV/c^2)	Decay	Branching fraction/per cent
	N*(1470)	$\frac{1}{2}^+$	1460	260	Nπ	55
					N$\pi\pi$	45
	N*(1518)	$\frac{3}{2}^-$	1515	115	Nπ	50
					N$\pi\pi$	50
	N*(1550)	$\frac{1}{2}^-$	1525	80	Nπ	35
					Nη	65
	N*(1680)	$\frac{5}{2}^-$	1675	145	Nπ	45
N*					N$\pi\pi$	55
$I = \frac{1}{2}$	N*(1688)	$\frac{5}{2}^+$	1690	125	Nπ	60
					N$\pi\pi$	40
	N*(1710)	$\frac{1}{2}^-$	1715	280	Nπ	65
	N*(1750)	$\frac{1}{2}^+$	1785	405	Nπ	34
	N*(2190)	$\frac{7}{2}^-$	2190	300	Nπ	35
	N*(2650)	?	2650	360	Nπ	
	N*(3030)	?	3030	400	Nπ	
	Δ (1236)	$\frac{3}{2}^+$	1236	120	Nπ	100
	Δ (1640)	$\frac{1}{2}^-$	1630	160	Nπ	25
					N$\pi\pi$	75
	Δ (1690)	$\frac{3}{2}^-$	1670	225	Nπ	15
Δ	Δ (1910)	$\frac{5}{2}^+$	1880	250	Nπ	20
$I = \frac{3}{2}$	Δ (1950)	$\frac{7}{2}^+$	1940	210	Nπ	40
					Δ (1236) π	$\sim$ 50
	Δ (2420)	$\frac{11}{2}^+$?	2420	310	Nπ	11
					N$\pi\pi$	> 20
	Δ (2850)	?	2850	400	Nπ	
	Δ (3230)	?	3230	440	Nπ	

Only the principal decay modes are given. In certain cases although the elasticity or proportion of decay into the elastic channel is known the nature of the inelastic modes is not known. Certain resonance candidates which seem doubtful have been omitted.

9.7 Resonances with $B = 1$ and $S = -1$

The study of these resonances has in general been by different means from the strangeness-zero particles. There are several reasons why this is so:

(a) For strangeness-zero resonances access can be had to both the $I = \frac{3}{2}$ and $I = \frac{1}{2}$ states by pion–nucleon elastic scattering. This has meant that high statistics elastic or charge-exchange scattering experiments, using counters, have been possible. For this reason formation experiments of high precision have been made yielding the data for detailed phase-shift analyses.

(b) For $|S| = 1$ resonances one requires in formation experiments an incident K-meson beam. Also, in production experiments the cross-sections for formation of final states containing strange particles are so much higher for incident kaons than for other incident projectiles that most of the work has been done with K-meson beams.

Since separated beams of K-mesons (see Chapter 1) came later than pion beams, and are always much weaker, formation experiments are more difficult. More important, however, these resonances have in general large branching ratios into channels other than KN. Indeed, since the threshold for decay into KN is about 1440 MeV/c^2, this channel is not accessible to the lower mass $S = -1$ resonances.

These factors have led to most of the study of the strange-particle resonances being carried out in bubble chambers, both for production and formation experiments. In general a separated beam of K-mesons is allowed to enter the chamber. In a formation experiment all channels will be studied as a function of incident momentum, batches of photographs being taken at 20–100 MeV/c intervals. In a production experiment, mass distributions may be studied at a single momentum. In either case the total number of photographs required is large, frequently exceeding half a million in a single production experiment, and ten to fifty thousand per momentum interval in formation.

We will illustrate the methods of study of the strangeness -1 resonances known as Y* by two examples, the first being the first Y* discovered, and being a simple case, and the second involving a more elaborate analysis.

The Y* (1385) is copiously produced in K$^-$p interactions over a wide range of incident momenta. The simplest process leading to production of this resonance is

$$K^-p \rightarrow \Lambda \pi^+ \pi^-, \qquad\qquad \textbf{9.5}$$

which appears in a bubble chamber as an event with two charged prongs and, if the Λ decays via pπ^-, a neutral vee. Application of energy and momentum conservation at the production and Λ-decay vertices, combined with observation of the track densities, can usually distinguish this process from the reaction of similar topology $K^-p \to K^0p\pi^-$. A Dalitz plot for the reaction **9.5** in which $m^2(\Lambda\pi^+)$ is plotted against $m^2(\Lambda\pi^-)$ shows two bands at masses of 1382 MeV/c^2 corresponding to formation of the so-called Y_1^* (1385). (Although the best mass known for this particle is 1382 MeV/c^2, it is still conventionally known by the mass from earlier measurements.)

The subscript indicates that the i-spin of this particle is 1. We note that since $B = +1$ and $S = -1$, then $Q = I_3 = \pm 1$ for the $Y^{*\pm}$ (1385). No higher charge manifestations of the Y^* have been found, so that we can write $I = 1$. The i-spins of the π- and Λ-decay products are 1 and 0 and we should expect the three charge states and decay modes

$$Y^{*+} \to \Lambda\pi^+, \qquad Y^{*0} \to \Lambda\pi^0, \qquad Y^{*-} \to \Lambda\pi^-.$$

The decay $Y^* \to \Sigma\pi$ is also possible, but the branching ratio for this decay is found to be only 10 ± 3 per cent.

The spin of the Y_1^* (1385) has been measured to be $\frac{3}{2}$ from a study of the decay angular distribution. The most useful distribution is that of the cosine of the angle between the normal to the production plane and the direction of one or other of the decay particles from the Y^*, as measured in the Y^* centre-of-mass. Any anisotropy in this distribution indicates a polarization of the Y^*, since correlation information is carried from the production to the decay vertex, but the details of the spin analysis will not be treated here.

The analysis of the Y^* (1520) (Watson, Ferro-Luzzi and Tripp, 1963) characteristics illustrates a number of important principles, and this experiment is rather a classic of its kind. We shall consider a formation experiment with K^- mesons on protons, the incident momenta ranging from 250–520 MeV/c. The studies were made using a liquid-hydrogen bubble chamber. We note first that at a mass of 1520 MeV/c^2 the resonance has open to it KN, $\Sigma\pi$, $\Lambda\pi$, $\Lambda\pi\pi$, and even $\Sigma\pi\pi$, decay channels of which the last is rather near threshold and will be neglected. We will consider therefore the processes:

$$K^-p \to \begin{cases} K^-p & \text{(a)} \\ \bar{K}^0n & \text{(b)} \\ \Sigma^{\pm}\pi^{\mp} & \text{(c)} \\ \Sigma^0\pi^0 & \text{(d)} \\ \Lambda\pi^0 & \text{(e)} \\ \Lambda\pi^+\pi^- & \text{(f)} \\ \Lambda\pi^0\pi^0 & \text{(g).} \end{cases}$$

The cross-sections, as a function of incident kaon momentum, for the above processes (a)–(f) are shown in Figure 70. From this data we can determine the i-spin of the Y^* (1520) (corresponding to incident K-meson momentum 390 MeV/c). The i-spin of the final states may be obtained with the aid of the appropriate Clebsch–Gordan coefficients. For the kaon–nucleon final states we have in the

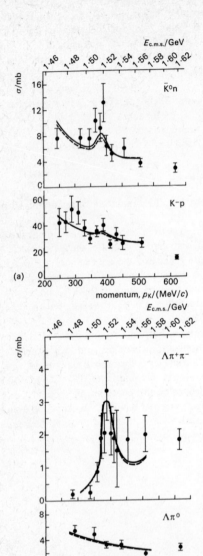
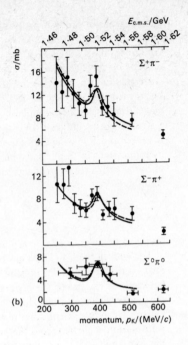

Figure 70 Cross-sections as a function of momentum for
(a) K^-p charge exchange and elastic scattering
(b) $\Sigma^+\pi^-$, $\Sigma^-\pi^+$ and $\Sigma^0\pi^0$ production,
(c) $\Lambda\pi^+\pi^-$ and $\Lambda\pi^0$ production.
The solid line corresponds to the best fit of all cross-sections, angular distributions, and polarizations for negative $KN\Sigma$ parity, while the dashed lines correspond to the best fit for positive $KN\Sigma$ parity

usual notation, where the first term in the brackets refers to I_3 for the nucleon and the second to I_3 for the kaon,

$$\psi_{I=0} = \sqrt{\tfrac{1}{2}}(\tfrac{1}{2}, -\tfrac{1}{2}) - \sqrt{\tfrac{1}{2}}(-\tfrac{1}{2}, \tfrac{1}{2}),$$
$$\psi_{I=1} = \sqrt{\tfrac{1}{2}}(\tfrac{1}{2}, -\tfrac{1}{2}) + \sqrt{\tfrac{1}{2}}(-\tfrac{1}{2}, \tfrac{1}{2}),$$

so that
$$\phi(p, K^-) = \sqrt{\tfrac{1}{2}}(\psi_{I=1} + \psi_{I=0}),$$
$$\phi(n, \bar{K}^0) = \sqrt{\tfrac{1}{2}}(\psi_{I=1} - \psi_{I=0}).$$

Also the $\Lambda\pi^0$ state is a pure $I = 1$ state, while the $\Sigma^0\pi^0$ and $\Lambda\pi^0\pi^0$ states formed in this process are pure $I = 0$ states. The student can readily verify these statements and write the appropriate wave functions to show that the $\Lambda\pi^+\pi^-$ and $\Sigma^\pm\pi^\mp$ states are $I = 0$, $I = 1$ mixtures. It is clear from the data that the rather clear presence of the resonance in the $\Sigma^0\pi^0$ and $\Lambda\pi^+\pi^-$ states, and its absence in $\Lambda\pi^0$ fix the i-spin as zero.

We can make some further quantitative checks on this conclusion. If we consider the $\Sigma\pi$ channels we have ratios

	$\Sigma^+\pi^-$	$\Sigma^0\pi^0$	$\Sigma^-\pi^+$
$I = 0$	1	1	1
$I = 1$	1	0	1

For the $\Lambda\pi\pi$ states we have

	$\Lambda\pi^+\pi^-$	$\Lambda\pi^0\pi^0$
$I = 0$	2	1
$I = 1$	1	0

The $\Lambda\pi^0\pi^0$ channel should account for most of the unfitted V-zero events at these low momenta and the experimental ratio for resonance decay in these modes is in fact found to be $\sim\tfrac{1}{2}$.

Finally we find that in *production* experiments of the kind

$$K^-p \to Y^*(1520) + \text{pions},$$

where there is no restriction on the Y^* charge, the resonance is still produced only in the neutral state.

We can now deduce the resonance spin, with some degree of certainty, from qualitative deductions from the differential cross-sections for the elastic and charge-exchange scattering, which are shown in Figure 71. Although the number of events at any energy is relatively small (~ 160 charge-exchange events at 390 MeV/c) the data quite clearly allows certain useful conclusions. In both the elastic and charge-exchange processes the 390 MeV/c distributions show a strong $\cos^2\theta$ dependence, which is absent above and below this momentum. Since no term higher than $\cos^{2J}\theta$ can occur, where J is the resonance spin, we see that $J > \tfrac{1}{2}$. The spin must of course be half-integral, since decay is into for instance KN. Also there is no need for $\cos^4\theta$ in any distribution so that $J \geqslant \tfrac{5}{2}$ is unlikely, although not absolutely excluded. If $J = \tfrac{3}{2}$ then we could have either a $p_{\tfrac{3}{2}}$ or a $d_{\tfrac{3}{2}}$ state, $(J^P = \tfrac{3}{2}^+ \text{ or } \tfrac{3}{2}^-)$. The angular distributions for decay of these states are the same

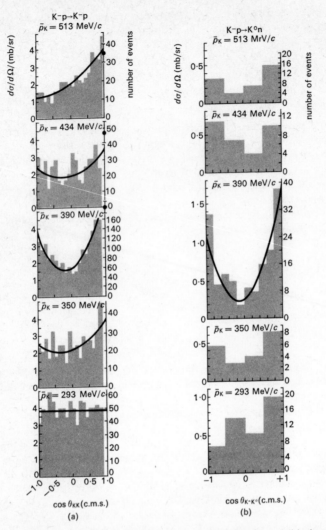

Figure 71 The differential cross-sections for (a) elastic and (b) charge exchange K⁻p scattering at various momenta in the vicinity of Y_0^* (1520) which is centred at 394 MeV/c with a half-width at half-maximum, corresponding to ±21 MeV/c K⁻ laboratory momentum (Watson, Ferro-Luzzi and Tripp, 1963)

(this is the well known Minami ambiguity). However it turns out that in this case the ambiguity can be resolved by considering the way in which the interference with the background changes as a function of momentum.

In order to do this fully we require to be able to write down the relations between the coefficients of the terms in the angular distribution

$$\frac{d\sigma}{d\Omega} \propto \sum_n A_n \cos^n \theta$$

and the amplitudes for scattering in the various states. This is beyond the scope of the present text so that we must make some unsubstantiated assertions. At incident momenta less than 250 MeV/c the angular distributions are isotropic, consistent with pure s-wave scattering, as one might expect at low momenta. At resonance the large $\cos^2\theta$ term is present, but the odd-power $\cos\theta$ interference term is not large. In this region we expect the continued existence of the s-wave background plus the resonance. However the amplitude–coefficient relationships referred to above show that s–p_3 interference leads to a large $\cos\theta$ term while the interference itself leads to no $\cos^2\theta$ term, although the p_3 alone will still yield such a term. On the other hand as s–d_3 interference develops it gives a fast increase in the $\cos^2\theta$ term, a decrease in the isotropic term and no $\cos\theta$ term; these are all in agreement with the observations, which are indeed quantitatively quite well fitted by a d_3 resonance ($J^P = \frac{3}{2}^-$) with an s-wave background.

An effort has been made to gain acceptance of a nomenclature for the Y* ($S = -1$) resonances in which $I = 0$ Y*s are known as, for instance, Λ (1520), Λ (1820) etc., while the $I = 1$ Y*s are referred to as Σ (1385), Σ (1660) etc. The presently known Y*s and their properties are given in Table 9.

Table 9 Baryon resonances with strangeness -1

	Resonance	J^P	Mass/ (MeV/c^2)	Width (Γ)/ (MeV/c^2)	Decay	Branching fraction/per cent
	Λ (1405)	$\frac{1}{2}^-$	1405	40	$\Sigma\pi$	100
	Λ (1520)	$\frac{3}{2}^-$	1519	16	$N\bar{K}$	45
					$\Sigma\pi$	45
					$\Lambda\pi\pi$	10
					$\Lambda\gamma$	0·9
	Λ (1670)	$\frac{1}{2}^-$	1670	25	$N\bar{K}$	14
					$\Lambda\eta$	33
					$\Sigma\pi$	45
	Λ (1700)	$\frac{3}{2}^-$	1690	40	$N\bar{K}$	25
Λ					$\Sigma\pi$	35
$I = 0$					$\Lambda\pi\pi$	20
					$\Sigma\pi\pi$	20

Table 9 Baryon resonances with strangeness -1 (*continued*)

Resonance	J^P	Mass/ (MeV/c^2)	Width (Γ)/ (MeV/c^2)	Decay	Branching fraction/per cent
Λ (1815)	$\frac{5}{2}^+$	1815	75	$N\bar{K}$	65
				$\Sigma\pi$	11
				Σ (1385) π	9
Λ (1830)	$\frac{5}{2}^-$	1830	80	$N\bar{K}$	10
				$\Sigma\pi$	35
Λ (2100)	$\frac{7}{2}^-$	2100	140	$N\bar{K}$	30
				$\Sigma\pi$	4
Λ (2350)	?	2350	210	$N\bar{K}$	
Σ (1385)	$\frac{3}{2}^+$	1382	36	$\Lambda\pi$	90
				$\Sigma\pi$	10
Σ (1615)	?	1615	65	$\Lambda\pi$	dominant
Σ (1660)	$\frac{3}{2}^-$	1660	50	Λ (1405) π	dominant
Σ (1700)	?	1700	110	$\Lambda\pi$	dominant
Σ (1765)	$\frac{5}{2}^-$	1765	100	$N\bar{K}$	46
				$\Lambda\pi$	16
				Λ (1520) π	15
				Σ (1385) π	15
Σ (1915)	$\frac{5}{2}^+$	1905	60	$N\bar{K}$	10
				$\Lambda\pi$	5
Σ (2030)	$\frac{7}{2}^+$	2030	120	$N\bar{K}$	10
				$\Lambda\pi$	35
				$\Sigma\pi$	10
Σ (2250)	?	2250	200	$N\bar{K}$	
Σ (2455)	?	2455	120	$N\bar{K}$	
Σ (2595)	?	2595	140	$N\bar{K}$	

(The Σ, $I = 1$ group label spans the Σ rows.)

9.8 Resonances with $B = 1$ and $S = -2$

Having seen that there exist resonances with $B = 1$, and $S = 0, -1$ and knowing also of the existence of the Ξ-particles with $B = 1$ and $S = -2$ it should be no surprise to find that there are also 'excited states' of the Ξ-particle or Ξ^* resonances. Those we might expect to decay via $\Xi\pi$, $\Lambda\bar{K}$, $\Sigma\bar{K}$ or even more complicated cascades. It is clear that, due to the comparative rarity of Ξ-particles or reactions involving Λ and K or Σ and K in the final state, the study of Ξ^* is always beset by the problem of statistics.

Nevertheless a number of Ξ^* resonances have been found. As an example we take the Ξ^* (1530), which was found in production in the process (Pjerrou *et al.*, 1962; Bertanza *et al.*, 1962).

$$K^-p \rightarrow \Xi^*K$$
$$\searrow \Xi\pi.$$

The ratio of Ξ^* decays to $\Xi^0\pi^-$, $\Xi^-\pi^0$, $\Xi^-\pi^+$ and $\Xi^0\pi^0$ gives the i-spin

$$\frac{\Xi^0\pi^-}{\Xi^-\pi^0} = \frac{\Xi^-\pi^+}{\Xi^0\pi^0} = \begin{cases} 2 \text{ for } I_{\Xi^*} = \frac{1}{2} \\ \frac{1}{2} \text{ for } I_{\Xi^*} = \frac{3}{2}. \end{cases}$$

The result strongly favours $I = \frac{1}{2}$.

The decay angular distributions indicate $J^P = \frac{3}{2}^+$ (or possibly $\frac{5}{2}^-$). $J^P = \frac{3}{2}^+$ allows this resonance to fit well into one of the multiplets of SU(3), which will be discussed in the following chapter.

An unusual feature of this resonance is its narrow width of $7 \cdot 3 \pm 1 \cdot 7$ MeV/c^2. The presently known Ξ^* particles are listed in Table 10.

Table 10 Baryon resonances with strangeness -2

Resonance	J^P	Mass/ (MeV/c^2)	Width (Γ)/ (MeV/c^2)	Decay	Branching fraction/per cent
Ξ (1530)	$\frac{3}{2}^+$	1530	7	$\Xi\pi$	100
Ξ (1820)	?	1820	20	$\Lambda\bar{K}$	60
				Ξ (1530) π	10
				$\Sigma\bar{K}$	30
Ξ (1930)	?	1930	110	$\Xi\pi$	
Ξ (2030)	?	2030	50	$\Lambda\bar{K}$	50
				$\Sigma\bar{K}$	50

9.9 Meson resonances: techniques

The study of the meson resonances has uncovered a rich world of particles of varied properties. All the early particles were discovered in production experiments in bubble chambers. Formation experiments for meson resonances require the production of an s-channel state, with baryon number zero, so that only through $\bar{p}p$ collisions can the meson resonances manifest themselves in formation. This implies that only meson resonances with masses greater than $2m_p$ can be studied in this way. So far the results of such experiments have not clearly revealed any new meson resonances.

An alternative technique to the bubble chamber has however yielded some very important results in this field. This depends on the use of arrangements generally known as missing-mass spectrometers which consist of combinations of counters, and frequently also spark chambers and analysis magnets. We shall describe an example of a successful missing-mass spectrometer used at CERN (Blieden et al., 1965).

The principle of this instrument is that one uses a beam of, say, π^- mesons, of well-defined momentum, incident on a hydrogen target and observes a recoil proton

$$\pi^-p \rightarrow pX^-.$$

X^- in this case represents all the other particles produced in the reaction which will together have a net negative charge. In the simplest arrangements the proton is identified and its vector momentum measured. Energy and momentum conservation then fix the total energy and momentum of the combination X^- and thus its effective mass can be established from

$$m_{X^-}^2 = (E-E_p)^2 - (\mathbf{p}_{\pi^-} - \mathbf{p}_p)^2, \qquad\qquad \textbf{9.6}$$

where E is the total laboratory energy, E_p and $\mathbf{p}_p$ the proton total energy and momentum and $\mathbf{p}_{\pi^-}$ the incident-pion momentum. If the X^- recoiling against the proton is a particle of well-defined mass, then a peak will be observed in $m_{X^-}^2$, while for two or more uncorrelated particles making up X^- we expect a smoothly varying background. In the earliest version of the CERN instrument (Figure 72)

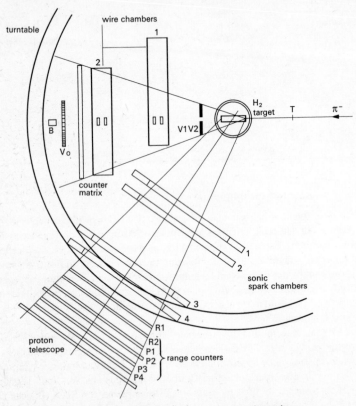

Figure 72 Diagram of the CERN missing-mass spectrometer.
The mass of X^-, in $\pi^- p \to X^- p$, is determined from the angle of the proton as measured in the proton-telescope spark chambers. The decay products from X^- are measured in the wire chambers (Kienzle, 1968)

the 'Jacobian peak' in the differential cross-section with respect to proton momentum was exploited. If we write **9.6** in the form

$$m_X^2 = (E - E_p)^2 - p_{\pi^-}^2 - p_p^2 + 2p_{\pi^-}\, p_p \cos\theta$$

and select protons by a rough measurement of p_p but work in the region where $dm_X/dp_p \simeq 0$, then m_X is a function only of the laboratory angle of the proton.

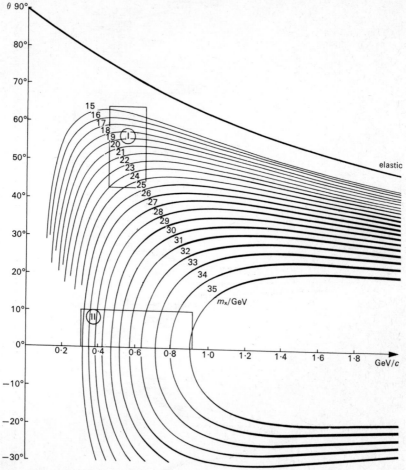

Figure 73 Kinematics of the process $\pi p \to Xp$ for incident pions of 12 GeV/c. The x-axis gives the proton momentum and the y-axis the proton angle in the laboratory. The curves give the equivalent mass for X. The missing-mass spectrometer was operated in region 1, where $dm_X/dp_p \simeq 0$ and m_X is a function only of θ_p. The CERN boson spectrometer (CBS) is operated in region 2 where m_X is a function only of p_p, which is measured by magnetic analysis (Kienzle, 1968)

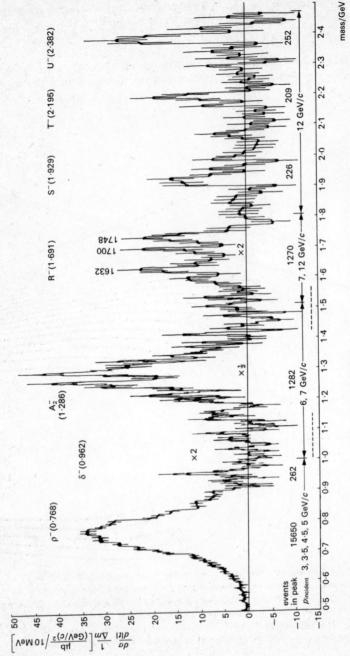

Figure 74 Compiled spectrum of bosons X^- from the reaction $\pi^- p \to p X^-$ observed in the CERN missing-mass spectrometer. The incident pion momenta are indicated below each peak. The spectrum has been scaled up by a factor of two in the region of the δ (900–1000 MeV) and R (1530–1810 MeV), and scaled down by a factor of two in the region of the A_2 (1150–1530 MeV). The bin size is 5 MeV for $0.90 < m < 1.00$ GeV, 7.5 MeV for $1.00 < m < 1.42$ GeV and 10 MeV for the rest of the spectrum. A smooth background has been subtracted throughout (Focacci et al., 1966)

The kinematics of the process is illustrated in Figure 73. The time of flight between an incident pion triggering the counter T and the proton pulses in the proton telescope, combined with the proton range, serve to identify the proton and give a broad measurement of its momentum. The angle of the proton however was determined accurately by means of the acoustic spark chambers in the telescope. The missing-mass spectrum resulting from a compilation of runs with the CERN 'MMS' is shown in Figure 74 and shows a number of peaks, some of which corres-

boson spectrometer layout 1968 (schematic)

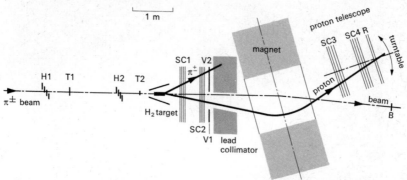

Figure 75 Schematic layout of the CERN boson spectrometer (CBS). Protons emerging near 0° are recorded in the spark chambers SC1, 2, 3, 4 and momentum analysed in the wide-gap magnet. Associated π-mesons fire scintillation counters V1 or V2 and are recorded in SC1, 2. H1 and H2 are counter hodoscopes giving the incident beam position. The beam–particle interaction is confirmed by anticoincidence of counter B. R is also a counter, and the time of flight between T2 and R, identifies the proton. A photograph of the equipment is shown in Figure 76 (Benz *et al.*, 1968)

pond to meson resonances observed in production experiments with bubble chambers and some of which have so far been seen only in the MMS.

A later instrument, also used at CERN and known as the CERN boson spectrometer (CBS), works in a different kinematic region also shown in Figure 73. The boson spectrometer is shown diagrammatically in Figure 75 and a photograph of the equipment is shown in Figure 76. In this case recoil protons emerging in a cone near 0° are momentum-analysed to approximately 1 per cent by a magnetic spectrometer, consisting of a wide-gap magnet and two pairs of large wire spark chambers. The time of flight of the protons also measures the momentum. Like the MMS, the whole arrangement can be revolved, in this case around the magnet centre, to study different mass regions. The physical scale of these devices can be judged by noting that the CBS wire spark chambers have dimensions 1.5×1.5 m^2. In both instruments the pulses were handled on-line by a computer, so that all data

Figure 76 The CERN boson spectrometer showing the two sets of spark chambers on either side of the large-aperture magnet (Kienzle, 1968)

collection was fully automatic. Other varieties of missing-mass spectrometer have been used in a number of experiments.

The principal advantage of the missing-mass spectrometer is the ability to collect very large numbers of events fairly rapidly. For instance the spectrum shown in Figure 74 results from $\sim 180\,000$ events (background has been subtracted). The resolution of these devices is apparently comparable with that of currently operating bubble chambers being 10–16 MeV/c^2 in the mass region $\sim 1 \cdot 3$ GeV/c^2 in the CBS. However, as we shall see, the MMS seems to detect fine structure not clearly seen in bubble-chamber experiments. The obvious disadvantage is that the information obtained about a meson resonance is limited, being confined to the mass and width in the simplest instruments. In the CERN instruments the number, and directions, of the charged decay products from the meson resonance were also measured in some of the runs, though in general it is true that much more precise information concerning the decay is available from bubble chambers.

9.10   Meson resonances: quantum numbers, J, I, P, C

The strongly interacting mesons are bosons, that is integral-spin particles, for which the baryon number is zero. Thus for mesons $Q = I_3 + \frac{1}{2}S$ and Y, the hyper-charge, $= S$. The mesons thus have integral i-spin if $S = 0$ and half-integral i-spin if $S = \pm 1$, as exemplified in the pions and kaons which we have already studied.

For the strongly decaying resonances we expect conservation of I as well as J, so that we can often draw conclusions about the quantum numbers of a meson resonance decaying into two particles from the nature of the final state. For decay into two spinless particles, the generalized Pauli principle (section 3.7) demands that if the particles are identical, e.g. $\pi^0 \pi^0$, then the relative orbital angular momentum l must be even. If the two decay products are not identical, but members of the same i-spin multiplet then:

$(l + I)$ must be even,

or in other words l and I must be even or odd together.

The above deductions from the generalized Pauli principle follow from the fact that the *parity* of the space wave function for two members of the same multiplet is $(-1)^l$. If J is the resonance spin, then particles having parity $(-1)^J$ are said to have natural spin–parity and those having parity $(-1)^{J+1}$ unnatural spin–parity. Decay into two pseudoscalar mesons is only possible for natural-parity resonances. In this connection we note that, unlike the baryons, meson particle and anti-particle have the *same* parity.

We have already used the *charge-conjugation* operator C in studying the K^0, $\bar{K}^0$ system (8.2). For a state to be an eigenfunction of C it must be electrically neutral, since in changing particle to antiparticle C reverses the charge. Equally, B and S must also be zero. In the decay of the K_1^0 we saw that the operations C and P were equivalent. This is similarly true for decay into K^+K^- and $K^0\bar{K}^0$ so that for these decays I, J, P and C must be all even or all odd. Also, as we have seen for the π^0 meson (2.8), decay to $\gamma\gamma$ implies even parity under charge conjugation.

We can also draw some simple conclusions for decay into K_1^0 and K_2^0 combinations. From our previous analysis of the K^0, $\bar{K}^0$ system it is clear that

$$CP|K_1^0K_1^0\rangle = CP|K_2^0K_2^0\rangle = (-1)^J,$$
$$CP|K_1^0K_2^0\rangle = -(-1)^J.$$

However, since C and P are equivalent for these systems they must be even under the CP operation, so that for K^0 decay

even-spin resonances can decay only to $K_1^0K_1^0$ or $K_2^0K_2^0$,

odd-spin resonances can decay only to $K_1^0K_2^0$.

9.11   **Meson resonances: G-parity**

When several conservation laws operate for the same system it is sometimes possible to obtain new quantum numbers and selection rules by combining the original ones. The new conservation law may reveal features not evident in the originals.

We recall that i-spin invariance holds good only for strong interactions. Charge-conjugation invariance holds for strong interaction and probably also for electromagnetic interactions. Only for strong interactions can we combine charge conjugation and i-spin invariance to obtain a new selection rule.

If the charge of a system is not zero it cannot be an eigenfunction of C. However if $B = S = 0$ the effect of C is simply to reverse the charge. Thus a system with $B = S = 0$ will be an eigenfunction of the combined operators CR where R is the charge inversion operator. The R-operator can be written in terms of the i-spin operators in that to invert the charge one needs to invert I_3, that is rotate the system by π about the I_2 axis. Thus $R = e^{i\pi I_2}$. The G-operator is

$$G = CR = Ce^{i\pi I_2}.$$

We may study the effect of the C-, R- and G-operators on π-mesons by writing the wave functions for the pions in terms of the i-spin components of the boson field. In these terms we write

$$\psi_{\pi^+} = \frac{1}{\sqrt{2}}(\phi_1 - i\phi_2),$$
$$\psi_{\pi^-} = \frac{1}{\sqrt{2}}(\phi_1 + i\phi_2),$$
$$\psi_{\pi^0} = \phi_3,$$

where ϕ_1, ϕ_2, ϕ_3 are the I_1, I_2, I_3 components of the field. Although we cannot prove these relations within the scope of the present text we may note at least that they satisfy the requirements concerning antiparticles and complex conjugation, discussed in connection with the K^0, $\bar{K}^0$ system.

When R is applied to the ϕs we see that, since it produces a rotation about the I_2 axis, ϕ_2 remains unchanged while the others reverse sign:

$$R\phi_1 = -\phi_1, \qquad R\phi_2 = \phi_2, \qquad R\phi_3 = -\phi_3.$$

Since $C(\pi^+) = \pi^-$ and $C(\pi^0) = \pi^0$, we see that

$$C\phi_1 = \phi_1, \qquad C\phi_2 = -\phi_2, \qquad C\phi_3 = \phi_3.$$

Thus $\quad G\phi_1 = -\phi_1, \qquad G\phi_2 = -\phi_2, \qquad G\phi_3 = -\phi_3,$

and $G(\pi) = -\pi$. Thus the pion wave functions are eigenfunctions of G with eigenvalue or G-parity -1.

We can derive the G-parity of an i-spin multiplet by considering its neutral member, since the G-parity is the same for all members of the multiplet. This may be seen to be so from the following argument. We can construct operators from the i-spin operators which will produce transformations between the members of an i-spin multiplet (cf. the raising and lowering operators in the quantum theory of angular momentum). G however may be shown to commute with all the i-spin operators $[G, \mathbf{I}] = 0$. Now suppose that we have a given G-parity for the neutral member of a multiplet, e.g. $G(\pi^0) = -1(\pi^0)$, and that we make an operator I^+ from I_1, I_2, I_3 which has the effect of increasing the value of the third component by one unit so that

$$I^+(\pi^0) = x\pi^+.$$

Now apply the operator G

$$
\begin{aligned}
G(\pi^+) &= G[\tfrac{1}{x}] \, I^+(\pi^0) \\
&= [\tfrac{1}{x}] \, I^+ G(\pi^0) \\
&= [\tfrac{1}{x}] \, (-1) I^+(\pi^0) \\
&= -1(\pi^+),
\end{aligned}
$$

where we have used the fact that G commutes with the i-spin operators from which I^+ is constructed.

The fact that the G-parity of a multiplet can be determined from that of its neutral member allows us to derive a useful relation between G-parity and i-spin. As with ordinary angular-momentum states we can represent the isotopic-spin wave function for a system, or particle, by a spherical harmonic $Y_I^{I_3}(\cos \theta)$, where θ is the angle which the i-spin vector makes with the I_3 axis. For the neutral member of a multiplet, with $B = S = 0$, I_3 is also zero so that the isotopic-spin wave function can be written $Y_I^0(\cos \theta)$. The R-operator (rotation by π about the I_2 axis) changes θ to $\theta + \pi$, which results in multiplying the wave function by $(-1)^I$. Thus for non-strange mesons

$$G = C(-1)^I.$$

If the meson decays *strongly* into $\pi\pi$ or $K\bar{K}$, then the C-parity is given by $(-1)^l$ and for this case the G-parity is given by

$$G = (-1)^{l+I}. \qquad \textbf{9.7}$$

9.12 Meson resonances: spin and parity

In the following chapter we shall see that a number of the mesons and meson resonances fall into multiplets of nine members (nonets), all of which have the same

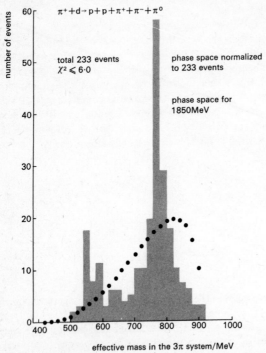

Figure 77 The first observation of the η-meson. The plot shows the effective mass of $\pi^+\pi^-\pi^0$ from the reaction $\pi^+ d \to pp\pi^+\pi^-\pi^0$. The upper peak corresponds to the ω-meson and the lower to the η-meson. The smooth points correspond to the calculated phase-space distribution for the total average (p, 3π) c.m.s. energy (1850 MeV) (Pevsner *et al.*, 1961)

spin and parity, and that these multiplets are accounted for by the 'SU(3)' symmetry group. It will thus be convenient for us to discuss the experimental evidence concerning the meson resonances by looking first at the *pseudo-scalar meson nonet* ($J^P = 0^-$) and then in turn at the *vector mesons* ($J^P = 1^-$) and tensor mesons ($J^P = 2^+$).

9.12.1   Pseudo-scalar mesons

We have already studied the isotopic-spin triplet of pions and the K^+, K^0 and $\bar{K}^0$, K^- i-spin doublets, all of which have $J^P = 0^-$. The other members of the nonet are the η and either the η′(X^0) or E-mesons.

In Figure 77 is shown the effective mass distribution of the $\pi^+ \pi^- \pi^0$ combination in the reaction in which the η was first observed,

$$\pi^+ d \rightarrow pp\pi^+ \pi^- \pi^0,$$

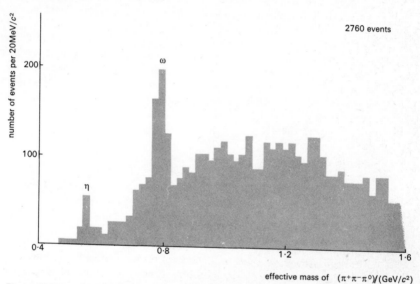

Figure 78   The effective mass of $(\pi^+ \pi^- \pi^0)$ in the reaction $K^+ p \rightarrow K^0 \pi^+ \pi^+ \pi^- \pi^0 p$ at 10 GeV/c. The peaks are due to formation of the η- and ω-resonances

at a momentum of 1·23 GeV/c. Instead of the smooth phase-space distribution the spectrum shows two sharp peaks at masses of 550 and 780 MeV/c^2. The upper of these is known as the ω-meson and the lower as the η-meson. Both of these resonances have also been seen in many other reactions. Figure 78 shows the distribution of $m(\pi^+\pi^-\pi^0)$ in the reaction

$$K^+p \rightarrow K^0\pi^+\pi^+\pi^-\pi^0 p.$$

No charged η has ever been observed, so that the i-spin is zero. The η is also found to have neutral decay modes

$$\eta \rightarrow \gamma\gamma,$$
$$\eta \rightarrow \pi^0\pi^0\pi^0,$$

which account for 71 per cent of all decays. The $\pi^+\pi^-\pi^0$ mode has a branching ratio of 23 per cent and the only other sizeable channel is

$$\eta \rightarrow \pi^+\pi^-\gamma \quad (5\cdot5 \text{ per cent}).$$

The γ-modes are clearly electromagnetic decays and the fact that these have an appreciable branching fraction suggests that the strong-decay modes must be inhibited by some conservation law and that the width should be $\lesssim 1$ keV, characteristic of electromagnetic processes.

The decay to $\gamma\gamma$ limits the possible spin to 0 or 2 (see section 2.8 concerning the spin of the π^0). For even spin–parity the η could decay strongly to two pions, so that the parity must be odd. The distribution in the Dalitz plot (cf. τ-decay) is fitted by $J^P = 0^-$. Since the η must be odd under CP (cf. K_1^0, K_2^0 discussion, 8.2) it is even under C (as can also be inferred from the $\gamma\gamma$ decay) and the G-parity is given by

$$G = +(-1)^0 = +.$$

But the G-parity of the three-pion final state is $(-1)^3 = -$, so that G-parity is not conserved in the decay. This accounts for the inhibition of the three-pion modes.

An upper limit of $0\cdot9$ MeV/c^2 has been set for the width by a time-of-flight experiment (Jones *et al.*, 1966). In this experiment the process

$$\pi^-p \rightarrow nZ^0$$

was studied near the η-threshold with a pion beam of well-defined momentum. The neutrons were detected by an array of plastic scintillation counters and their time of flight measured by a delayed coincidence between the incident pion and the neutron pulses. A veto counter ensured that no charged particles left the hydrogen target. A unique mass for Z^0 is reflected as a peak in the neutron time-of-flight spectrum, which showed sharp peaks for the Z^0 corresponding to γ, π^0 and η^0 production. The shape of the excitation function just above threshold was studied by varying the incident-pion momentum, and this shape is particularly sensitive to the η-width. The results of this measurement allowed the width to be established as $<0\cdot9$ MeV/c^2.

The η' meson is also narrow ($\lesssim 5$ MeV), with mass 958 MeV and decay modes

$$\eta' \rightarrow \pi^+\pi^-\eta,$$
$$\eta' \rightarrow \pi^+\pi^-\gamma.$$

The quantum numbers are most likely

$$IJ^{PCG} = 0, 0^{-++}.$$

The E-meson appears to be most readily produced in antiproton annihilations, in hydrogen, in reactions like

$$\bar{p}p \to K^0_1 K^{\pm} \pi^{\mp} \pi^- \pi^+,$$

where it appears as a peak in the mass of the $(K\bar{K}\pi)$ combination. Detailed analysis suggests $IJ^{PCG} = 0, 0^{-++}$ although $J^P = 1^+$ is not ruled out. The properties of the pseudo-scalar mesons are summarized in Table 11.

Table 11 The pseudo-scalar nonet

Meson	Mass/ (MeV/c^2)	Width/ (MeV/c^2)	I J^{PCG}	Decays	Branching fraction/per cent
$\pi^{\pm}$	140			$\mu\nu$	100
π^0	135		1 0^{-+-}	$\gamma\gamma$	100
$K^{\pm}$	494			see 5.2	
$K^0\bar{K}^0$	498		$\frac{1}{2}$ 0^-	and 5.4	
η	549	2·63* ±0·64 keV/c^2	0 0^{-++}	$\pi^+\pi^-\pi^0$	23
				$\pi^+\pi^-\gamma$	6
				$\gamma\gamma$	38
				$\pi^0\pi^0\pi^0$	29
$\eta'(X^0)$	958	<4 MeV/c^2	0 0^{-++}	$\eta\pi^+\pi^-$	71
				$\pi^+\pi^-\gamma$	22
				$\gamma\gamma$	6

*Inferred from a Primakoff-effect measurement (Bemporad et al., 1967)

Only the well-established decay channels are listed. In this nonet the members and their quantum numbers are all well established with the possible exception of the spin and parity of the η'.

9.12.2 Vector mesons

Unlike the pseudo-scalars, the vector meson nonet includes no semi-stable ($\tau \sim 10^{-10}$) particles but has only strongly-decaying members. These are the

$$\rho^{\pm 0}, \quad \omega^0, \quad K^{*\pm 0}, \quad \bar{K}^{*0}, \quad \phi^0.$$

It is not appropriate to this text to describe in detail the evidence and the arguments used in determining the quantum numbers and other properties of all these resonances. We shall rather present some examples and list the properties in Table 12.

Table 12 The vector nonet

Meson	Mass/ (MeV/c^2)	Width/ (MeV/c^2)	I	J^{PCG}	Decays	Branching fraction/per cent
ρ	765	125 ± 20	1	1^{--+}	$\pi\pi$	~ 100
ω	783	$12 \cdot 6 \pm 1 \cdot 1$	0	1^{---}	$\pi^+\pi^-\pi^0$	~ 90
					$\pi^0\gamma$	9
$K^*(890)$	891	$49 \cdot 7 \pm 1 \cdot 1$	$\frac{1}{2}$	1^-	$K\pi$	100
$\bar{K}^*(890)$						
ϕ	1020	$3 \cdot 7 \pm 0 \cdot 6$	0	1^{---}	K^+K^-	48
					$K_1^0 K_2^0$	33
					$\pi^+\pi^-\pi^0$	19

Possible very small branching fraction decays are not listed.

In Figures 77 and 78 we have already seen examples of ω-production in the mass combination $(\pi^+\pi^-\pi^0)$. No charged ω has been seen, so we expect $I = 0$. $G = -1$, so $C = -1$ (strong decay) and L for the di-pion $(\pi^+\pi^-)$ is odd (Figure 79). For the i-spin wave function we get $I = 0$ by combining an $I = 1$, $\pi^+\pi^-$ di-pion with $I = 1$ for the π^0. For the di-pion the $I = 1$ system is antisymmetric (for $I_3 = 0$, $I = 1$ we have

$$\frac{1}{\sqrt{2}}(\pi^+\pi^- - \pi^-\pi^+)$$

– see Clebsch–Gordan coefficients), so for the three-pion system the i-spin wave function is also antisymmetric. Since overall symmetry is required by Bose–Einstein statistics, the space part of the wave function must also be antisymmetric. Since C is odd so is L. The assignments are given in Table 13.

Table 13

J^P for ω	L	l	Parity of space w.f.	Dependence on momenta
0^-	1	1	$+$	$\mathbf{p.q}$ – zero on medians
1^+	1	0	$-$	$\mathbf{q}$ – zero at $q = 0$
1^-	1	1	$+$	$\mathbf{p} \times \mathbf{q}$ – zero on boundary ($\mathbf{p} \| \mathbf{q}$)

As before $\mathbf{q}$ is the momentum of either π in the di-pion c.m.s. and $\mathbf{p}$ is the momentum of the π^0 relative to the di-pion in the three-pion c.m.s. The dependencies in the last column lead to rather distinctive distributions in the Dalitz plot, which allow determination of $J^P = -1$.

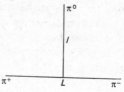

Figure 79

The ρ-*meson* is a broad resonance ($\Gamma = 125 \pm 20 \, \mathrm{MeV}/c^2$, $m = 765 \pm 10 \, \mathrm{MeV}/c^2$) found to decay into $\pi^+\pi^0$, $\pi^+\pi^-$, $\pi^-\pi^0$, and is thus seen to be likely to have i-spin one. This hypothesis is confirmed by measurement of the ratio of ρ^-/ρ^0 production in the reactions

$$\pi^-p \to \begin{cases} \pi^0\pi^-p \\ \pi^+\pi^-n. \end{cases}$$

In these two reactions, and in the process

$$\pi^+p \to \pi^+\pi^0p,$$

ρ-production is copious, the momentum transfer to the proton is small and the other features of the reactions suggest that they proceed via one-pion exchange (Figure 80). In this case the ratio of ρ^-/ρ^0 production in the first two reactions will be (see Appendix B for C–G coefficients) $\frac{1}{2}$, 0, $\frac{2}{9}$ for $I(\rho) = 1$, 0 and 2 respectively. Experimentally, the ratio is found to be 0·5.

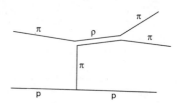

Figure 80

Figure 80 illustrates a method by which we can (indirectly) study π–π scattering at the upper vertex, although not 'on the mass shell' (i.e. with energy and momentum conserved) since the exchanged pion is virtual. However an apparently successful recipe exists which allows extrapolation to the actual physical situation. Many analyses of ρ-production using these reactions have studied the distribution of $\cos\theta$, where θ is the $\pi\pi$ scattering angle in the c.m.s. of the di-pion. For the simplest model, with $J(\rho) = 1$, this distribution should be pure $\cos^2\theta$. Initial- and final-state effects may modify the distribution by addition of an isotropic term to give

$$A + B\cos^2\theta,$$

but the distribution should still be symmetrical about 90°. For the charged ρ, the angular distributions below and above the resonance are asymmetric, as would be expected if the resonance amplitude interferes with a background which is itself changing only slowly. At resonance the angular distribution becomes symmetric (Figure 81). For the ρ⁰, however, the asymmetry persists *in* the resonance and does not change sign passing through the resonance.

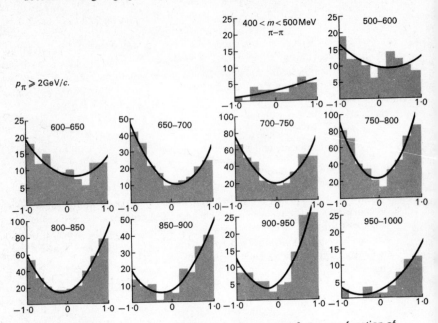

Figure 81 The $\pi^-\pi^-$ scattering angle (see text) in $\pi^- p \rightarrow \pi^0\pi^- p$, as a function of $(\pi^0\pi^-)$ mass. Only events with four-momentum squared (Δ^2) less than $5(m_\pi)^2$ have been used to ensure 'peripheral' pion production, corresponding to the diagram of Figure 80. The forward–backward asymmetry is seen to disappear at 750 MeV/c^2, corresponding to the ρ-mass (Walker *et al.*, 1967)

The most plausible explanation of this asymmetry occurring only in the neutral state is that there may exist an $I = 0$ interaction, almost in phase with the ρ-amplitude, which may also probably resonate. This resonance has been sought in a number of experiments but has not so far been found. Since an $I = 1$ resonance cannot decay into $\pi^0\pi^0$ (consider the appropriate C–G coefficients), while an $I = 0$ resonance *can* decay in this way, the $\pi^0\pi^0$ channel is a good one in which to seek the $I = 0$ resonance (variously known as the ε^0 or S^0). However, study of this channel is difficult since it involves detection of the γ-rays from the π^0 decays.

The K*(890) resonances appear strongly in almost every reaction in which they can be produced. For instance Figure 67 (p. 167) shows the $K\pi$ mass spectrum from the process

$$K^+p \rightarrow K^0p\pi^+$$

for incident kaons of momentum 10 GeV/c. The distribution shows peaks at 890 and 1400 MeV/c^2. By studying reactions such as

$$K^+p \rightarrow K^{*+}p \begin{cases} \rightarrow K^0\pi^+p \\ \rightarrow K^+\pi^0p, \end{cases}$$

at incident kaon momentum high enough to produce the K^*, but not so high that the secondary K^+ and p are frequently indistinguishable in the fit or in ionization, one may determine the ratio

$$\frac{K^{*+} \rightarrow K^0\pi^+}{K^{*+} \rightarrow K^+\pi^0}.$$

This ratio is found to be 0·5, as expected for $I = \frac{1}{2}$ ($I = \frac{3}{2}$ gives a ratio of 2). The decay angular distributions fix J as 1 so the parity is -1. The strangeness is $+1$ for K^{*+}, K^{*0}, and -1 for $\bar{K}^{*0}$ and K^{*-}, so G is not a good quantum number.

The ϕ-*meson* is observed to decay into $K\bar{K}$ only in neutral states and no charged ϕ has ever been observed, though if it existed such an object could readily be produced in for instance

$$K^-p \rightarrow \Sigma^+K^0K^-.$$

The mass of the ϕ is 1020 MeV/c^2 and it is found to be very narrow. An example of ϕs produced in the processes

$$K^+p \rightarrow K^0\bar{K}^0K^+p \quad (K^0_1K^0_2K^+p)$$
$$\rightarrow K^+K^-K^+p,$$

is shown in Figure 82. The Q-value for the decay into K^+K^- is only 32 MeV, which means that for ϕs formed near threshold the laboratory kinetic energies of the K^+ and K^- are small. Measurement of the range of K^+ and K^- stopped in a bubble chamber in such an experiment can give precise values for the energies of these particles, and hence for the ϕ-mass. By this means the ϕ-width has been established as $3\cdot7\pm0\cdot6$ MeV/c^2. The only satisfactory fit to the decay angular distribution is for $J^P = 1^-$. From the relation **9.7** we see $G = -1$.

9.12.3 *The tensor nonet*

The tensor nonet is a group of mesons all of spin–parity 2^+. The assignment of particles to this group is rather less certain than for the pseudo-scalar and vector nonets. Its members are generally taken to be the A_2, f, f^1 and K^* (1400) resonances.

The f-*meson* decays into two pions and is seen only in the neutral state. The ratio

$$\frac{f \rightarrow \pi^0\pi^0}{f \rightarrow \pi^+\pi^-}$$

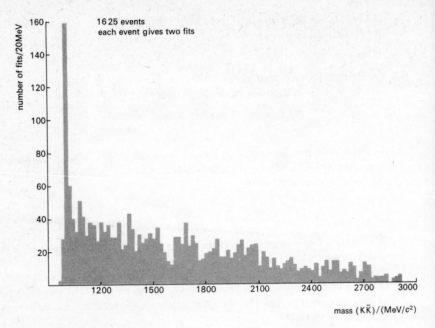

Figure 82 The effective mass distribution for $K\bar{K}$ from the reactions
$K^+p \rightarrow K^0K^-K^+p$ and $K^+p \rightarrow K^+K^-K^+p$, for incident mesons of momentum 10 GeV/c.
The sharp spike at 1019 MeV/c^2 is due to the ϕ-meson (Birmingham–Glasgow
10 GeV/c collaboration)

is in agreement with 0·5, so $I = 0$, $G = +1$, J^P must be 0^+, 2^+, 4^+, . . . and
$C = +1$. The distribution of $\cos \theta$ (θ is the $\pi\pi$ scattering angle, as in the discussion
of the ρ-meson) is symmetrically forward–backward peaked, so 0^+ is excluded
while 2^+ gives a reasonable fit. A convincing analysis of the spin–parity for particles
decaying to two pions is illustrated in Figure 83. Figure 83(a) shows the $\pi^+\pi^-$
mass distribution found in a study of the process

$$\pi^-p \rightarrow \pi^+\pi^-n$$

at 6 GeV/c (Crennell *et al.*, 1967). The mass distribution shows clear peaks corresponding to the ρ-meson and the f-meson (1264 GeV/c^2) and also a small peak
corresponding to a meson of even higher mass (1650 GeV/c^2) known as the
g-meson. The $\pi^-\pi^-$ scattering angular distribution for events with four-momentum
transfer less than 1·0 (GeV/c)^2 was fitted to the distribution

$$A_n P_n \cos \theta$$

and the even coefficients are plotted as a function of $m(\pi^+\pi^-)$ in Figure 83(b),
(c), (d). The highest coefficient to be expected is A_{2J} and we see that for the ρ-
region ($J = 1$) only A_2 is found, while for the f-region both A_2 and A_4 are

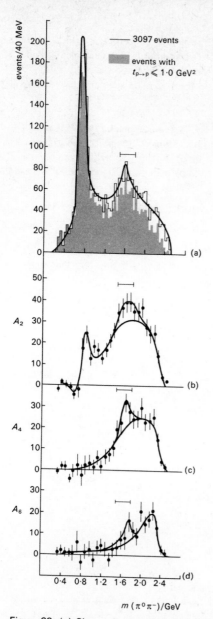

Figure 83 (a) Shows the effective mass distribution for $\pi^+\pi^-$ from the reaction $\pi^-p \to \pi^-\pi^+n$ at 6 GeV/c. The peaks due to the ρ^0, f^0 and g-mesons are visible. (b), (c) and (d) show the variation with $m(\pi^+\pi^-)$ of the coefficients A_n when the $\pi^-\pi^-$ scattering distribution is fitted to A_nP_n (cos θ). The presence of A_4 and absence of A_6 for the f^0 mass suggests $J = 2$ (Crennell *et al.*, 1967)

substantial while A_6 is absent, indicating $J = 2$. Although A_6 is also present for the g, the decay is very asymmetric (substantial odd-A coefficients) and it is not possible to determine J for the g from this distribution.

The f^1 *meson* appears as a peak in the $K\bar{K}$ mass spectrum in reactions like

$$K^-p \rightarrow \Lambda K_1^0 K_1^0$$

at a mass of 1514 ± 5 MeV/c^2. The $K_1^0 K_1^0$ mode means that this meson is even under charge conjugation. A good illustration of this point was presented by the discoverers of the f^1 (Barnes *et al.*, 1965). For any $K^0\bar{K}^0$ system one can derive the proportion of events with one (N_1) or with two (N_2) visible K^0 decays under the assumptions:

(a) that there is no resonance so that the K^0 and $\bar{K}^0$ decay uncorrelated,

(b) that the K^0, $\bar{K}^0$ arise from the decay of a resonance with $C = +1$ and

(c) that the K^0, $\bar{K}^0$ arise from the decay of a resonance with $C = -1$.

The result is shown as a function of $m^2(K^0\bar{K}^0)$ in Figure 84. The reader should check the expected values of the ratio

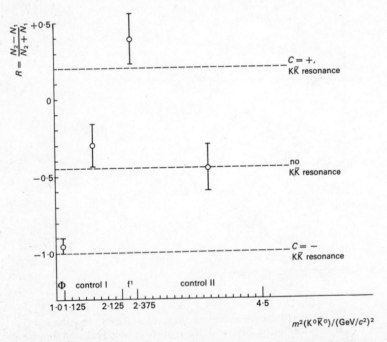

Figure 84 The ratio R (see text) as a function of m^2 ($K^0\bar{K}^0$) for the process $K^-p \rightarrow \Lambda K^0\bar{K}^0$. In the f^1 region the ratio reaches ~ 0.2, as expected for a resonance with $C = +1$ (Barnes *et al.*, 1965)

$$R = \frac{N_2 - N_1}{N_2 + N_1},$$

as an exercise.

The K^* (1400) meson has already been mentioned, in passing, in the discussion of the K^* (890) and Figure 67. Its decay angular distributions establish fairly clearly that $J^P = 2^+$ and the branching to different charge states gives unambiguously that $I = \frac{1}{2}$.

The A_2 *meson*, the isotopic-spin triplet of the nonet, presents one of the most novel and puzzling features in the meson sprectrum.

The A_2 has been seen in many bubble-chamber experiments as a peak at 1320 MeV/c^2 in the $\pi\rho$ mass distribution. It is also observed to decay into $K_1^0 K_1^0$ and $K^- K_1^0$. All the data is consistent with $I = 1$, $C = +1$. The $K_1^0 K_1^0$ mode requires $J^P = 0^+, 2^+, \ldots$. For $A_2 \to \pi\rho$, however, $J^P = 0^+$ is forbidden so that 2^+ is the lowest allowed assignment. The decay angular distributions, although not conclusive, also favour this assignment.

The A_2 meson is clearly seen in the CERN missing-mass spectrometer experiment (Figure 74) but in the kinematic region where the resolution is good ($\lesssim 16$ MeV/c^2), the peak resolves into a doublet (Figure 85), the two peaks A_2^L and A_2^H having masses 1289 and 1309 MeV/c^2. Some evidence for splitting of the A_2 peak has also come from bubble-chamber experiments for both the $\pi\rho$ mode and the $K_1^0 K^{\pm}$ decay, although there are also experiments where no splitting is seen. However, the evidence from the missing-mass spectrometer work appears to be conclusive regarding the existence of the splitting†.

The explanation of the splitting is currently a matter of speculation. Three possible explanations have been suggested:

(a) Two close Breit–Wigner resonances with the same strength and width ($\Gamma \simeq 22$ MeV/c^2) and probably also the same J^P value of 2^+.

(b) A 'dipole' resonance where instead of the usual Breit–Wigner formula one has a cross-section proportional to

$$\frac{\Gamma^2 (E - E_0)^2}{[(E - E_0)^2 + \frac{1}{4}\Gamma^2]^2}.$$

This would be the only known example of what is mathematically a more general class of resonance but the physical significance of which is not understood.

(c) A broad A_2 resonance plus another destructively interfering narrow resonance exactly at its centre.

Perhaps more likely than any of these proposals is the possibility that this 'fine structure' is due to some more fundamental reason not recognized in present theory.

The properties of the $J^P = 2^+$ mesons are summarized in Table 14.

† Missing-mass experiments, carried out at CERN and Brookhaven (1971) since this book was written, find no evidence for splitting of the A_2, so the experimental situation is not clear.

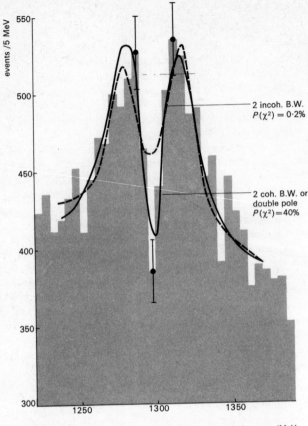

Figure 85 The sum of data from the CERN missing-mass spectrometer group in the region of the A_2 meson. The A_2 is seen to be split. The fits are for a sum of two incoherent Breit–Wigner amplitudes and for a double pole or dipole. The probability for the dipole fit (40 per cent) is acceptable while that for the incoherent fit is low (0–2 per cent) (Benz *et al.,* 1968)

9.12.4 *Other meson resonances*

There are a number of other meson resonances, some fairly well established and others less certain. Many of these resonances have problems associated with their quantum numbers and other properties, and lie very much in the field of presently active research the results of which are not sufficiently stable to include in a text of this kind. It is also likely that other meson resonances remain to be discovered.

Table 14 The tensor ($J^P = 2^+$) nonet

Meson	Mass/ (MeV/c^2)	Width/ (MeV/c^2)	I	J^{PCG}	Decays	Branching fraction/per cent
f	1264	150 ± 25	0	2^{+++}	$\pi\pi$	~ 100
A_2	1297†	91 ± 10†	1	2^{++-}	$\rho\pi$	dominant
					$K\bar{K}$	
					$\pi\eta$	
K^* (1400)	1422	90 ± 6	$\frac{1}{2}$	2^+	πK	51
					πK^* (890)	33
					ρK	11
					ωK	3
					ηK	2
f^1	1514	73 ± 23	0	2^{+++}	$K\bar{K}$	~ 80
					$\eta\pi\pi$	~ 20

†The problem of the split A_2 is discussed in the text.

For the full summary at any instant the student is referred to the so-called 'Rosenfeld Tables' published as UCRL 8030 and also, periodically, in *Reviews of Modern Physics*.

For the purpose of the classification and models discussed in the following chapter, however, we refer once more to the spectrum of negative mesons revealed by the CERN missing-mass spectrometer and shown in Figure 74. In addition to the ρ^- and A_2^-, discussed in detail in the text, we see there the δ^- (962), the R^-, a triple peak with masses 1632, 1700 and 1748 MeV/c^2, the S^- (1929), T^- (2195), U^- (2382). A resonance known as the g-meson, decaying into $\pi^-\pi^\delta$ and having mass ~ 1700 MeV/c^2, and very probably spin 3, has been seen in bubble-chamber experiments and probably corresponds to at least part of the R-complex seen in the missing-mass spectrometer.

Chapter 10
Classification of particles: SU(3) and the quark model

10.1 Introduction

The discovery of such a wealth of apparently 'elementary' particles stimulated new activity in the search for a *pattern* amongst them, as a first step towards the understanding of their nature. The discovery of such a pattern is analogous to, for instance, the discovery of the Rydberg formula in atomic spectroscopy. Only the Bohr atom finally provided an explanation of the formula, as hopefully may the quark model provide an explanation of the symmetry pattern of the elementary particles.

We have already become familiar with the limited symmetry of isotopic-spin multiplets. In that case we grouped together particles which were the same except for properties associated with the electric charge. The degeneracy of the multiplet is removed by the symmetry-breaking Coulomb interaction. Alternatively, we can regard the members of the multiplet as states linked by rotations in isotopic-spin space and we can define a group of rotation operators which enable us to step from one state to another.

The Coulomb interaction is not strong compared with the so called 'strong' interactions and the symmetry breaking to which it gives rise is small. For instance the masses of particles in the same isotopic-spin multiplet only differ by at most a few per cent. In order to *extend* the symmetry, to group larger numbers of particles together, we must recognize the existence of much stronger symmetry-breaking forces since the mass differences between, say, i-spin multiplets are substantial, even compared with the particle masses themselves. Nevertheless it turns out to be true that the symmetry, though broken, remains in many ways very useful.

10.2 Baryon and meson multiplets

In attempting to group together different i-spin multiplets we may seek to group particles having the same baryon number, spin and parity, but allow the strangeness (or equivalently the hypercharge) to vary within the larger multiplet. This hypothesis has the merit of success over others, such as allowing spin and parity to vary within a multiplet and demanding the same strangeness, which might *a priori* appear equally reasonable.

For i-spin multiplets, we may represent the members of the multiplet as points spaced at unit intervals on the I_3 axis. The raising and lowering operators allow steps to the right and left along the axis. For instance, for the Δ (1236) multiplet we have

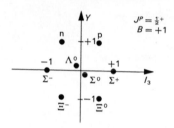

Figure 86

In order to extend the classification to include other i-spin multiplets of the same J^P and B we need to move from a one- to a two-dimensional diagram, where the axes are I_3 and Y. We first take the $J^P = \frac{1}{2}^+$ baryons, of which eight were known when the classification was first proposed. The original $\frac{1}{2}^+$ octet consists of p, n, Σ^+, Σ^0, Σ^-, Λ^0, Ξ^0 and Ξ^-; two i-spin doublets, a triplet and a singlet. The diagram for this octet is shown in Figure 87.

Figure 87

The octet forms a hexagonal pattern on the Y–I_3 diagram and it is clear that in order to make transitions between states we need operators which effect diagonal steps as well as the step operators within the i-spin multiplet.

Many other particles also readily fall into similar patterns. In particular for the pseudo-scalar, vector and tensor mesons we find nonets which are shown on the Y–I_3 plot in Figure 88.

In the meson nonets we notice that, unlike the baryon case, particle and anti-particle appear in the same multiplet, since for both particle and antiparticle B is the same.

10.3 Symmetry groups

Having noted the existence of these multiplets we may seek a classification to describe them. The SU(3) group was proposed for this purpose in 1961 by Gell-Mann and independently by Neeman.

Some of the formal properties of groups are summarized below though the reader wishing to pursue this aspect of the subject more deeply is recommended to consult one of the several excellent texts such as *Lie Groups for Pedestrians* (Lipkin, 1965). In this section we use some of these properties to bring out the principal physical ideas.

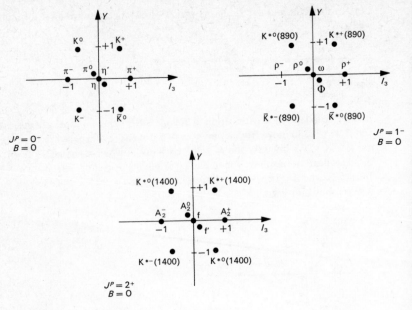

Figure 88

The idea of examining the symmetry of any system by 'rotating' it in the appropriate space is a familiar one. We accomplish this by means of 'rotation' operators, in the most general sense of the term rotation. The set of rotation operators is said to form a *symmetry group*. For instance, on functions of geometric coordinates we have the translation, space-rotation and inversion groups. These operators act on the wave functions.

The most familiar example to compare with the operators we shall require is the space-rotation group. These operators act on the angular coordinates of the wave function $\psi(\theta_1, \theta_2, \theta_3) = \psi(\theta)$. They have the form

$$R(\theta) = e^{i\theta \cdot J}.$$

Here the θ are the Euler angles and J the angular momentum. These operators form an infinite group.

Now if we have a limited space of states such that under the operators of the group each state acted on by an operator of the group transforms into another state in this space, then we have an *invariant subspace* of states. For such a set of states, an operator of the group cannot connect a state within the set to a state outside. For instance the set of functions $Y_{0,0}$, $Y_{1,1}$, $Y_{1,0}$, $Y_{1,-1}$ form a four-dimensional invariant subspace of the rotation group. The subspace is said to be *irreducible* if it contains no smaller subspace. Thus, pursuing the above example, we see that:

$Y_{0,0}$ is an irreducible singlet,

$Y_{1,1}$, $Y_{1,0}$, $Y_{1,-1}$ is an irreducible triplet.

The conventional way of expressing this division is to write

$$(Y_{0,0}, Y_{1,1}, Y_{1,0}, Y_{1,-1}) = Y_{0,0} \oplus (Y_{1,1}, Y_{1,0}, Y_{1,-1}).$$

An irreducible invariant subspace is known as a *multiplet* of the operator group.

A general theorem can be proved which states that if the Hamiltonian for the system commutes with the operators of the symmetry group then the multiplets are sets of degenerate eigenstates of the system. We know that for the elementary particles the states are not degenerate, so that the multiplet is *split*. This is due to additional terms in the Hamiltonian which do *not* commute with the group operators. An example of terms which cause splitting for the rotation group is for instance the **L.S** coupling which splits the levels in atomic spectra. The nature of the Hamiltonian for the elementary particles is not known. The success of the symmetry scheme and the splitting indicate the need for a 'strong' and a 'medium-strong' part to the Hamiltonian.

In the space-rotation group there are an infinite number of operators, but we can deal with this infinite group in a simple way because all the operators can be expressed in terms of only three basic operators J_1, J_2 and J_3. If $R(\theta)$ depends analytically on θ, at least near $\theta = 0$ and thus near $R(\theta) = 1$, then we can relate the Js to R by means of the relation

$$iJ_k = \left[\frac{\partial R(\theta)}{\partial \theta_k} \right]_{\theta = 0}.$$

A *Lie group* is a continuous group composed of operators $U(\boldsymbol{\alpha}) = U(\alpha_1, \ldots, \alpha_n)$ which depends analytically on all the n parameters and for which $U(\mathbf{0}) = 1$. As in the particular case above we can extract basic operators

$$G_i = \left. \frac{\partial U}{\partial \alpha_i} \right|_{\boldsymbol{\alpha} = \mathbf{0}},$$

which are called the *generators* of the group but which are not themselves members of the group. The *order* of the group is given by the number of the generators.

We have already seen that the larger multiplets for particles *contain* the i-spin multiplets. Suppose, therefore, that we enlarge a Lie group by adding more coordinates for the group to transform. In this case we have to add more operators to the group.

An idea of fundamental importance in the present application of group theory is that of the *basic or elementary multiplet*. For all Lie groups there exists such an elementary multiplet, and from this can be built up more complicated multiplets by taking repeated 'products' between the basic multiplet and itself. A simple example from the rotation group is the triplet deuterium ground state consisting of spin-$\frac{1}{2}$ neutron and proton. The spin-$\frac{1}{2}$ particles are members of two-member multiplets or doublets of the rotational group for angular momentum $\frac{1}{2}$ (the generators of which are the Pauli spin matrices). The product of two such doublets then yields triplet and singlet multiplets according to $2 \otimes 2 = 3 \oplus 1$. This kind of decomposition of the product of two multiplets is a process which is familiar under another guise in angular momentum theory as the Clebsch–Gordan expansion

$$Y_{l_1 m_1}(\theta, \phi) \, Y_{l_2 m_2}(\theta, \phi) = \sum_{L=|l_1-l_2|}^{l_1+l_2} \sum_{M=-L}^{+L} C_{m_1 m_2 M l_1 l_2 L} Y_{LM}(\theta, \phi).$$

The form of the decomposition or the values of the Clebsch–Gordan coefficients must be worked out from the commutation relations of the generators.

Thus for the Simple Unitary (Lie) group of order 3 (SU(3)) the elementary multiplet is a triplet, and the multiplet products can be shown to decompose as, for instance,

$$3 \otimes 3 \otimes 3 = 10 \oplus 8 \oplus 8 \oplus 1.$$

10.4 The SU(3) classification and the quark model

For SU(2) there are only two basic states, that is the basic multiplet is a doublet such as the p, n i-spin multiplet. In order to include states with non-zero hypercharge, it is necessary to move to a larger group of order 3, such as the group SU(3) for which the basic multiplet is a triplet. The state to be added to the SU(2) doublet is then a state with non-zero hypercharge and $I = 0$. The strangeness and isotopic-spin quantum numbers of the triplet are then the same as p, n, Λ.

An early model in which it was indeed proposed that all the known particles were built from the physical proton, neutron and lambda particles was proposed by Sakata in 1956. However this model quite rapidly ran into some problems, such as the parity of the Σ, which was on this model composed of N$\bar{\text{N}}\Lambda$ (e.g. $\Sigma^+ \equiv \text{p}\bar{\text{n}}\Lambda$) in mutual s-states, so that $J^P = \frac{1}{2}^-$ and not $\frac{1}{2}^+$ as is experimentally found to be the case. Although this proposal has been abandoned, the basic SU(3) multiplet is still usually referred to as 'p', 'n', 'Λ' and their antiparticles, sometimes known as 'sakatons' or more commonly known as 'quarks'.

It has often been stressed that the SU(3) classification does not depend on the actual existence of particles corresponding to the basic multiplet. We shall return to the question of existence of the quarks below. For the present we merely use a basic triplet of states, equivalent under the strong interaction, to construct the pattern of the observed particles.

In order to find the patterns we require the set of operators forming the group SU(3) which will transform the basic triplet states. For SU(3) the operators may be represented by a group of unimodular, unitary 3×3 matrices in a similar way to the Pauli spin matrices for SU(2). There are $3 \times 3 - 1$ such independent traceless matrices. However, the proper treatment of these eight operators requires a more detailed knowledge of group theory than can be presented here.

We rather follow the treatment of Lipkin (1965) to derive the operators in terms of bilinear products of the creation and annihilation operators for 'p', 'n' and 'Λ'. These operators are denoted a_p^+, a_n^+, a_Λ^+ and a_p, a_n, a_Λ respectively.

The technique is best illustrated by application first to the simplest case of the i-spin doublet p, n. Only bilinear products which do not change the baryon number can be permitted, so that in this case there are four such products:

$a_p^+ a_n$–changes n to p $\equiv$ step up operator $= \tau_+$,

$a_n^+ a_p$–changes p to n $\equiv$ step down operator $= \tau_-$,

$a_p a_p^+$–counts protons (i.e. the number of protons is given by the eigenvalue),

$a_n a_n^+$–counts neutrons.

In order to include states with $S \neq 0$ we include the a_Λ^+, a_Λ, obtaining

$a_\Lambda^+ a_p$–changes p to $\Lambda = C_-$,

$a_\Lambda^+ a_n$–changes n to $\Lambda = C_+$,

$a_p^+ a_\Lambda$–changes Λ to p $= B_+$,

$a_n^+ a_\Lambda$–changes Λ to n $= B_-$.

Together with the first two we thus have six step operators.

τ_+ and τ_- leave Y unchanged and change I_3 by ± 1,

B_+ and B_- change Y by $+1$ and I_3 by $\pm\frac{1}{2}$,

C_+ and C_- change Y by -1 and I_3 by $\pm\frac{1}{2}$

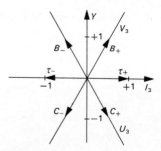

Figure 89

Such a set of step operators satisfies the rule that operation on any state transforms it into another state of the multiplet. These operators require a hexagonal-type symmetry for all the multiplet diagrams. This is due to the fact that instead of the single I_3 symmetry axis in SU(2) there are now three symmetry axes at 120° to each other.

Along the 'I-spin' axis the hypercharge is constant but the charge varies. The step operators for the basic multiplet change 'n' $\leftrightarrow$ 'p'. Along the $B_- - C_+$ axis the charge Q remains constant, Y varies and the step operators change 'n' $\leftrightarrow$ 'Λ'. This is known as the 'U-spin' axis. If invariance under I_3 changes was exact then the properties of a system would be independent of its charge. If invariance under U-spin was true the properties would be independent of the hypercharge. The third axis, corresponding to 'p' $\leftrightarrow$ 'Λ' transformations, has no such simple interpretation. It is sometimes known as the V-spin axis.

As has already been mentioned, in SU(3) there are eight independent operators of which we have discussed the six step operators formed from bilinear combinations of the creation and annihilation operators for 'p', 'n', 'Λ'. It is convenient to choose the other two operators as

$$\tau_0 = \tfrac{1}{2}(a_p^+ a_p - a_n^+ a_n),$$
$$N = \tfrac{1}{3}(a_p^+ a_p + a_n^+ a_n - 2a_\Lambda^+ a_\Lambda).$$

By writing the charge, strangeness and baryon number operators which count these quantities as Q, S and B we see that

$$Q = a_p^+ a_p,$$
$$S = -a_\Lambda^+ a_\Lambda,$$
$$B = a_p^+ a_p + a_n^+ a_n + a_\Lambda^+ a_\Lambda,$$

so that using $Q = I_3 + \tfrac{1}{2}(B+S)$ we have that operation with τ_0 gives the I_3 eigenvalue. Also

$$N = \tfrac{1}{3}B + S.$$

None of these operators changes B, N changes S and thus Y.

For these eight operators the basic triplet and antitriplet are shown in Figure 90.

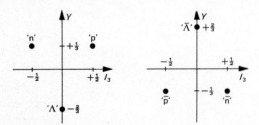

Figure 90

The operators transform the basic states into each other. By combining these basic triplets nine states are obtained, which can be generated from the vacuum by the products of creation and annihilation operators. This is shown in Figure 91, where we have omitted the states at the centre generated by the remaining three products or by operators such as τ_0, B and N. The nine states split into an octet plus a singlet. The singlet has the quantum numbers of the vacuum, while the octet also has two states with $I_3 = Y = 0$. As far as the SU(3) algebra is concerned these two octet states are degenerate, but we note that the four $Y = 0$ states must include the $I_3 = 0$ member of the i-spin triplet. Thus the two states at the origin have $I = 1$ and $I = 0$.

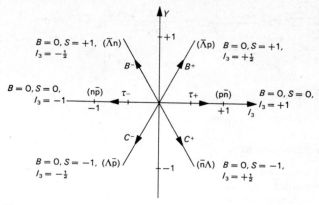

Figure 91

From this analysis we see that in the simplest model all the mesons ($B = 0$) can be constructed from quark–antiquark (q$\bar{\text{q}}$) pairs.

In a similar way the baryons and antibaryons may be constructed from three quark (qqq) or three antiquark ($\bar{\text{q}}\bar{\text{q}}\bar{\text{q}}$) combinations. We shall return to this aspect of the model in section 10.7.

A consequence of the nature of the step operators is that the multiplets are all hexagonal lattices in the Y–I_3 plane, having the following general properties:

(a) All diagrams are symmetrical about the Y-axis ($+I_3 \leftrightarrow -I_3$) corresponding to such transformations as n $\leftrightarrow$ p and also about axes corresponding to p $\leftrightarrow \Lambda$ and n $\leftrightarrow \Lambda$ transformations.

(b) The multiplicity of points at any coordinate increases by one at each 'ring' as one moves in from the boundary, until one arrives at a point, or triangle, inside which triangle it remains constant.

(c) Charge conjugation changes the sign of Y and I_3. Thus the charge-conjugate multiplet is the original reflected through the origin.

SU(3) gives, in addition to the multiplets we have already discussed, others, such as those formed by combining two octets or three triplets

$$8 \otimes 8 = 27 \oplus 10 \oplus \overline{10} \oplus 8 \oplus 8 \oplus 1$$
$$3 \otimes 3 \otimes 3 = 10 \oplus 8 \oplus 8 \oplus 1.$$

On the Y–I_3 diagram the decuplet and 27-plet have the forms shown in Figure 92.

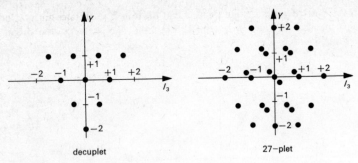

decuplet 27–plet

Figure 92

10.5 Mass formulae: baryons

The 'strong' interactions are by definition the same for all the particles in the multiplet. The electromagnetic interactions cause a small mass splitting along the I_3 axis. The electromagnetic interaction Hamiltonian does not commute with the group operators. This splitting removes the degeneracy along the I_3 axis.

The mass splitting between the different I-spin multiplets within the same SU(3) multiplet is postulated to be due to a 'medium strong' force. In terms of quarks this symmetry breaking along the Y- or S-axis may be accounted for by giving the Λ-quark a greater mass than the n- and p-quarks.

The basic assumption is thus that the 'strong' interaction is scalar in I-spin and U-spin. The electromagnetic interaction is then seen to be a scalar in U but to have a vector dependence on I_3,

$$m(I, I_3) = m_0(I) + x(I) I_3,$$

giving the charge splitting.

We may readily test the simplest similar assumption that the mass splitting between I-spin multiplets is scalar in I and vector in U:

$$m(U, U_3) = m_0(U) + y(U) U_3. \qquad \textbf{10.1}$$

Since $U_3 = Y - \tfrac{1}{2}Q$ and Q is a constant in a U-spin multiplet we have

$$m(U, U_3) = m_0(U) + Y. \qquad \textbf{10.2}$$

This result is directly applicable to the decuplet. From experiment were known nine baryons with $J^P = \tfrac{3}{2}^+$ which appeared to fit well into a decuplet pattern as shown in Figure 93.

At this time (1962), no particle was known for the bottom apex of the triangle. The SU(3) classification requires such a particle and the so-called 'Ω' was postulated to complete the decuplet. The properties of the Ω were readily predicted as

$$B = +1, \quad J^P = \tfrac{3}{2}^+, \quad Y = -2, \quad S = -3, \quad I = 0, \quad I_3 = 0, \quad Q = -1.$$

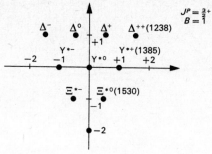

Figure 93

For this multiplet, where no space is occupied by more than one particle, the relation **10.2** gives an equal-mass spacing rule

$$m_{\Omega^-} - m_{\Xi^*(1530)} = m_{\Xi^*(1530)} - m_{Y^*(1385)} = m_{Y^*(1385)} - m_{\Delta(1238)}.$$

The best values for the latter two differences are

$$m_{\Xi^*(1530)} - m_{Y^*(1385)} = 147 \pm 1 \cdot 5 \text{ MeV}/c^2,$$
$$m_{Y^*(1385)} - m_{\Delta(1238)} = 146 \pm 1 \cdot 4 \text{ MeV}/c^2.$$

Thus the mass of the Ω^- is predicted to be

$$m_{\Xi^*(1530)} + 146 = 1675 \text{ MeV}/c^2.$$

With such a mass, the only $\Delta S = 1$ decays which are kinematically allowed are

$$\Omega^- \rightarrow \begin{cases} \Lambda K^- \\ \Xi^- \pi^0 \\ \Xi^0 \pi^-. \end{cases}$$

The discovery of the Ω^-, its decay by the above modes and the measurement of its mass, the best present value of which is $1672 \cdot 4 \pm 0 \cdot 6$ MeV/c^2, have already been described in 5.10. The number of Ω^-s seen so far is such that no information is available concerning its spin and parity. This prediction and its verification provide one of the triumphs of the SU(3) classification.

The derivation of the mass relation in the octet is slightly more involved due to the fact that there are two states at the origin. We have already seen that these are easily distinguished as I-spin triplet and singlet states, the mass of the Σ^0 linking it unambiguously with Σ^+ and Σ^-. In U-spin however we do not have such a separation and the $U_3 = 0$ member of the U-spin triplet will be a mixture of the Σ^0 and Λ states. For this purpose we shall use the raising and lowering operators for U-spin and I-spin, analogous to the J_+ and J_- of ordinary angular momentum. In Figure 94 we see that the state n is related to Λ, Σ^0 by the lowering operator U_-. Carrying over the coefficient from ordinary angular momentum theory we have

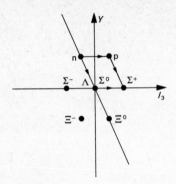

Figure 94

$$U_-|U, U_3\rangle = \sqrt{[U(U+1)-U_3(U_3-1)]}|U, U_3-1\rangle,$$

so that $\quad U_-|1, +1\rangle = \sqrt{2}|1, 0\rangle$.

If we write the $U = 0$ member of the U-spin triplet as a mixture of the Σ^0 and Λ states, we have

$$|U = 1, U_3 = 0\rangle = \alpha|\Sigma^0\rangle + \beta|\Lambda\rangle,$$

where α and β are to be determined.

Thus

$$U_-|n\rangle = \sqrt{2}[\alpha|\Sigma^0\rangle + \beta|\Lambda\rangle].$$

Now transform to the Σ^+ state by applying the operator I_+ so that

$$I_+U_-|n\rangle = \sqrt{2}\alpha\sqrt{[I(I+1)-I_3(I_3+1)]}|\Sigma^+\rangle,$$

where the Λ-term falls out. Thus

$$I_+U_-|n\rangle = 2\alpha|\Sigma^+\rangle.$$

However we can also reach the Σ^+ via the proton

$$U_-I_+|n\rangle = U_-|p\rangle = |\Sigma^+\rangle,$$

since U_- and I_+ must commute. Thus $\alpha = \frac{1}{2}$. Normalization demands that

$$|\alpha|^2+|\beta|^2 = 1,$$

so that $\beta = \pm\sqrt{\frac{3}{4}}$. We arbitrarily choose the positive sign for β and write

$$|U = 1, U_3 = 0\rangle = \tfrac{1}{2}|\Sigma^0\rangle + \sqrt{\tfrac{3}{4}}|\Lambda\rangle.$$

The U-spin triplet is then

$$|n\rangle, \quad \tfrac{1}{2}|\Sigma^0\rangle + \sqrt{\tfrac{3}{4}}|\Lambda\rangle, \quad |\Xi^0\rangle.$$

Applying the relation **10.1**, and squaring the coefficients to obtain the expectation values, we get

$$m_n - (\tfrac{1}{4}m_{\Sigma^0} + \tfrac{3}{4}m_\Lambda) = (\tfrac{1}{4}m_{\Sigma^0} + \tfrac{3}{4}m_\Lambda) - m_{\Xi^0}$$

or
$$\frac{m_n + m_{\Xi^0}}{2} = \frac{m_{\Sigma^0} + 3m_\Lambda}{4}, \qquad\qquad \textbf{10.3}$$

the famous Gell-Mann–Okubo mass formula.

This result is also rather well verified by experiment:

$$\frac{m_n + m_{\Xi^0}}{2} = 1127{\cdot}2 \pm 0{\cdot}4 \text{ MeV}/c^2,$$

$$\frac{m_{\Sigma^0} + 3m_\Lambda}{4} = 1134{\cdot}8 \pm 0{\cdot}1 \text{ MeV}/c^2.$$

In fact these mass relations are special cases of a more general mass formula

$$m = a + bY + c[I(I+1) - \tfrac{1}{4}Y^2].$$

10.6 Mass formulae: mesons

We might expect that the Gell-Mann–Okubo formula could be applied to the meson octets. For mesons the field equations always involve the squares of the masses, so that it is plausible that the formula here should be written with mass squared instead of mass as in the baryon case. A test of whether mass or mass squared is the correct quantity is complicated by the phenomenon of mixing which we shall discuss below.

If we then write the Gell-Mann–Okubo formula **10.3** in a general form for mesons where the subscripts $\tfrac{1}{2}$ and 1 refer to the I-spin and where m_8 is the mass of the I-spin zero member of the octet. Then

$$m_{\frac{1}{2}}^2 = \tfrac{1}{4}m_1^2 + \tfrac{3}{4}m_8^2, \qquad\qquad \textbf{10.4}$$

since for the mesons particle and antiparticle appear reflected about the I_3 axis.

We have already noted that the mesons appear to be grouped in nonets rather than octets. In fact the $I = 0$ member of the SU(3) octet has quantum numbers identical to those of the SU(3) singlet so that if there is SU(3) breaking we might expect mixing between these particles. We will denote the true SU(3), $I = 0$, $Y = 0$, octet and singlet particles as α_8 and α_0 respectively. However the observed physical particles may be (specific) mixtures of the α_8 and α_0 and the multiplet containing these mixtures will have nine members. The idea of mixing was proposed by Sakurai (1962) who proposed also a mixing parameter in the form of the 'mixing angle'. This parameterization makes it easy to preserve the normalization and has other formally satisfying features. The situation is very similar to that of the K^0, $\bar{K}^0$ particles. In that case the mixing was due to the weak interaction which breaks strangeness conservation while here the mixing is presumably due to the medium strong SU(3) breaking interactions.

Formally both the $K^0-\bar{K}^0$ and actet–singlet mixing can be treated rather similarly by means of the '*mass matrix*'. We shall not attempt here to develop the treatment of the mass matrix fully or rigorously, but merely introduce the idea and use it to obtain an expression for the mixing angle.

First note that we expect to obtain the mass, m_8, of the *pure octet state* from the expression **10.4**. For instance in the vector meson nonet we should have

$$m_8^2 = \tfrac{4}{3}m_{K^*(890)}^2 - \tfrac{1}{3}m_\rho^2 = 0 \cdot 863 (\text{GeV}/c^2)^2.$$

In this nonet the physical singlet particles are the ω and the ϕ for which the mass-squared values are $0 \cdot 610$ and $1 \cdot 035$ $(\text{GeV}/c^2)^2$ respectively.

Now we write the physical particle wave functions, β and β' say, in terms of the Sakurai mixing parameter

$$|\beta\rangle = |\alpha_0\rangle \cos\theta + |\alpha_8\rangle \sin\theta,$$
$$|\beta'\rangle = -|\alpha_0\rangle \sin\theta + |\alpha_8\rangle \cos\theta. \qquad \textbf{10.5}$$

For a stable stationary state, the solutions of the wave equation contain a factor $e^{-im_0 t}$, and the mass of the particle is the eigenvalue of the total Hamiltonian for the interaction. Thus

$$\langle\beta|H_0|\beta\rangle = m_\beta^2 \quad (\text{e.g. } m_\omega^2),$$
$$\langle\beta'|H_0|\beta'\rangle = m_{\beta'}^2. \qquad \textbf{10.6}$$

If there is no mixing then

$$\langle\beta|H_0|\beta'\rangle = \langle\beta'|H_0|\beta\rangle = 0,$$

but if some part of H_0 mixes the basic states this will no longer be true.

For the no-mixing case we could write a two-component time-dependent Schrödinger equation of the form

$$i\frac{d}{dt}\begin{bmatrix} \beta \\ \beta' \end{bmatrix} = \begin{bmatrix} m_\beta^2 & 0 \\ 0 & m_{\beta'}^2 \end{bmatrix} \begin{bmatrix} \beta \\ \beta' \end{bmatrix},$$

where in this case the *mass matrix* is diagonal.

If there is mixing the off-diagonal terms are no longer zero and we have

$$i\frac{d}{dt}\begin{bmatrix} \alpha_0 \\ \alpha_8 \end{bmatrix} = \begin{bmatrix} m_0^2 & m_{0,8}^2 \\ m_{8,0}^2 & m_8^2 \end{bmatrix} \begin{bmatrix} \alpha_0 \\ \alpha_8 \end{bmatrix}, \qquad \textbf{10.7}$$

where m_0 and m_8 are the masses of the *I*-spin zero members of the SU(3) singlet and octet multiplets and $m_{0,8}$ and $m_{8,0}$ are due to mixing. The mass matrices corresponding to the representation of the states in terms of the physical particles and the pure SU(3) states must satisfy the conditions that their traces and determinants are equal:

$$m_0^2 + m_8^2 = m_\beta^2 + m_{\beta'}^2,$$
$$m_0^2 m_8^2 - m_{0,8}^4 = m_\beta^2 m_{\beta'}^2.$$

($m_{0,8}$ has been taken *equal* to $m_{8,0}$).

The two representations must be linked by a rotation which is the operator

$$R = \begin{bmatrix} \cos\theta & -\sin\theta \\ \sin\theta & \cos\theta \end{bmatrix}.$$

Thus θ is the angle of rotation which will diagonalize the mass matrix

$$M = \begin{bmatrix} m_0^2 & m_{0,8}^2 \\ m_{8,0}^2 & m_8^2 \end{bmatrix}$$

of equation 10.7. Imposing this condition in the form

$$R^{-1}MR = \begin{bmatrix} m_\beta^2 & 0 \\ 0 & m_{\beta'}^2 \end{bmatrix}. \tag{10.8}$$

Straightforward algebra shows that the condition that the off-diagonal elements in RMR^{-1} be zero is

$$\tan 2\theta = \frac{m_{0,8}^2 + m_{8,0}^2}{m_8^2 - m_0^2}. \tag{10.9}$$

10.5 and 10.6 or consideration of the diagonal elements of 10.8 yield

$$\begin{aligned} m_\beta^2 &= m_0^2 \cos^2\theta + m_8^2 \sin^2\theta - (m_{0,8}^2 + m_{8,0}^2)\sin\theta\cos\theta, \\ m_{\beta'}^2 &= m_0^2 \sin^2\theta + m_8^2 \cos^2\theta + (m_{0,8}^2 + m_{8,0}^2)\sin\theta\cos\theta. \end{aligned} \tag{10.10}$$

We may eliminate $(m_{0,8}^2 + m_{8,0}^2)$ and m_0^2 from 10.9 and 10.10 to obtain an expression for the mixing angle in terms of m_8^2 (which can be measured). This expression can be written

$$\sin^2\theta = \frac{m_\beta^2 - m_8^2}{m_{\beta'}^2 - m_\beta^2}.$$

For the pseudo-scalar mesons we find $\theta = \pm 10.4°$ (mixing η and η'). For the vector mesons $\theta = \pm 39.9°$ (ω,) and for the tensor mesons $\theta = \pm 29.9°$, (f, f^1). Thus for the pseudo-scalar mesons the mixing is small and the physical η nearly obeys the Gell-Mann–Okubo formula, but for both the vector and tensor mesons the mixing is large and the physical particle masses do not obey the GMO relation.

Note that by introducing the new free parameter θ, the mass relation ceases to be a test of the theory or to be a predictive tool. There are other results involving θ which are open to independent test, such as the ratios of certain decay modes of the $I = 0$ mesons within the nonet and, on the basis of a particular quark interaction model, the ratios of production of say η and η' in certain reactions. Such results, although in general agreement with the mixing angles obtained from the masses, usually have large errors for θ while the model used in the production reaction analysis is also not well proved.

10.7   **Mesons and baryons constructed from quarks**

The quarks as building blocks for the mesons and baryons were introduced in 10.4. From Figure 90, and the relations between Y, B, S, Q and I_3, we can deduce

Table 15 Properties of the Quarks

	Y	I	I_3	B	S	Q
p	$+\frac{1}{3}$	$\frac{1}{2}$	$+\frac{1}{2}$	$\frac{1}{3}$	0	$\frac{2}{3}$
n	$+\frac{1}{3}$	$\frac{1}{2}$	$-\frac{1}{2}$	$\frac{1}{3}$	0	$-\frac{1}{3}$
Λ	$-\frac{2}{3}$	0	0	$\frac{1}{3}$	-1	$-\frac{1}{3}$
$\bar{\text{p}}$	$-\frac{1}{3}$	$\frac{1}{2}$	$-\frac{1}{2}$	$-\frac{1}{3}$	0	$-\frac{2}{3}$
$\bar{\text{n}}$	$-\frac{1}{3}$	$\frac{1}{2}$	$+\frac{1}{2}$	$-\frac{1}{3}$	0	$+\frac{1}{3}$
$\bar{\Lambda}$	$+\frac{2}{3}$	0	0	$-\frac{1}{3}$	$+1$	$+\frac{1}{3}$

the properties of the quarks, which are summarized in Table 15.

The simplest assumption is that all the quarks have spin $\frac{1}{2}$. Thus, for instance a K^+ meson will be constructed from a $p\bar{\Lambda}$ quark combination and a proton from ppn.

We have already seen that SU(3) allows singlets, octets, decuplets, antidecuplets, 27-plets and higher multiplets. However the quark model as described above is more restrictive. For mesons we expect only

$$3 \otimes \bar{3} = 8 \oplus 1,$$

i.e. singlets and octets, while for baryons we expect only

$$3 \otimes 3 \otimes 3 = 1 \oplus 8 \oplus 8 \oplus 10,$$

i.e. no antidecuplet or 27-plet or higher multiplet.

For instance consider the meson 27-plet shown in Figure 95. Some possible decay modes are shown for the 'far-out' states such as a doubly charged $S = +2$ meson (K^+K^+), a doubly charged ($I = \frac{3}{2}$) $S = +1$ meson (doubly charged

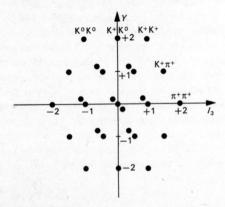

Figure 95

$K^* \rightarrow K^+\pi^+$) and a doubly charged $S = 0$ meson (decaying to $\pi^+\pi^+$ or $\pi^0\pi^+\pi^+$). In order to construct such an $S = +2$, $I_3 = +1$ meson we require four quarks, $\bar\Lambda\bar\Lambda pp$. Similarly the student may check that a strangeness $S = +1$ and charge-2 baryon must belong to a baryon 27-plet and be constructed from five quarks ($\bar\Lambda pppn$), while a $B = +1$, $S = +1$ singly charged particle must belong to an antidecuplet.

Some evidence of such states has been reported in formation experiments in the form of peaks in the total cross-section for K^+p scattering (pure $I = 1$) and in the total $I = 0$, K–nucleon scattering cross-section as deduced from the K^+p and K^+n (K^+d) cross-sections. However, the peak in the K^+p cross-section may probably be explained by the opening up of inelastic pion-production channels, while other possible explanations of the $I = 0$ peak have also been proposed. In addition a number of doubly charged peaks have been found in production experiments. These peaks have in general either found explanations as reflections of other resonances or have not been substantiated. Thus, at the present time, all well-established meson resonances fit into octets or singlets (nonets) and all the baryon resonances into singlets, octets or decuplets. It seems clear that 'far-out' or 'exotic' resonances are produced only weakly, if at all, and this result must be counted as evidence for the correctness of the quark model.

10.8 The orbital-excitation model for quarks

If the mesons are taken to consist of $q\bar q$ pairs, the q and $\bar q$ may rotate about each other with relative orbital angular momentum $l\hbar$. We may then have a series of states of increasing values of l. Such a model has been proposed by Dalitz.

Mesons built in this way must have some specific properties. The parity is given by $P = (-1)^{l+1}$ and the charge conjugation parity in the strangeness-zero case will be $C = (-1)^{l+s}$, where s is the spin of the $q\bar q$ pair. Thus for instance all natural parity states must have $C = P$.

The lowest $q\bar q$ states are then the $^3s_1$ and $^1s_0$ states, which have the appropriate J^{PC} for the vector and scalar nonets respectively. From the basic $^3s_1$ and $^1s_0$ nonets can be generated two series of rotational levels of increasing l. The triplet $^3s_1$ yield p-states $^3p_0$, $^3p_1$ and $^3p_2$, for which $P = +1$, and for the neutral states of which $C = +1$. The singlet p-state is $^1p_1$, with again positive parity, but negative under charge conjugation.

The four p-states may be rendered non-degenerate by spin–orbit SU(3) breaking forces proportional to $\mathbf{l.s}$. But since

$$\langle \mathbf{l.s} \rangle = \frac{j(j+1)-l(l+1)-s(s+1)}{2},$$

we have equal spacing in mass^2 for the p-levels

$$^3L_{L-1} \quad ^3L_L \quad ^1L_L \quad ^3L_{L+1}$$

Spacing $\propto$ $\leftarrow 2L \rightarrow \leftarrow 2 \rightarrow \leftarrow 2L \rightarrow$

For d-states the inner states are twice as close as the other spacings.

It is possible to assign some of the higher-mass mesons to such orbital-excitation multiplets. However the presently existing data is not sufficient to establish or to fault the model.

For the baryon three-quark system the orbital-excitation model is, of course, considerably more complicated.

10.9 Regge trajectories

SU(3) enables us to group together particles of the same spin, parity and baryon number. In the previous section we have also seen that the quark orbital-excitation model gives a relationship between multiplets corresponding to different angular-momentum states.

A more general relationship between multiplets of different spin is afforded by the so-called 'Regge trajectories'. These, although recently proposed and having basic features (e.g. linearity) the reasons for which are not understood, afford such a striking connection with experiment, adding a new dimension to particle classification, that they deserve mention here even though the Regge theory and its many applications are beyond the scope of this text.

In a study of the non-relativistic Schrödinger equation in a Yukawa-potential well, Regge examined the use of *complex* values of the angular-momentum quantum number J. With such a formalism the scattering amplitude can be analytically continued in the complex J-plane with the help of the solutions of the Schrödinger equation. In any scattering problem the situation can be described by what is called the S-matrix, which is the ratio between the incident and scattered amplitudes

$$S = \frac{A_1}{A_s}.$$

In the full development of the Regge theory these amplitudes are expressed in terms of J and the energy E^2, both of which are formally taken to be continuous complex variables

$$S(J, E^2) = \frac{A_1(J, E^2)}{A_s(J, E^2)}.$$

Thus S is a function in a two-dimensional complex space. A *Regge-pole* is a singularity in this space. As E^2 varies the pole will move through this space. If we look at the real-J–real-E^2 plane, the moving pole traces out a path which is often referred to loosely as the Regge trajectory. Such a plot is known as a Chew–Frautschi plot, and the detailed shape of the trajectory depends on the nature of the interaction potential.

When the trajectory passes through an integral J-value it gives rise to a pole of S on the J–E^2 plane which we may interpret as a bound state or resonance. Thus we may look on the trajectories on the Chew–Frautschi plot as lines linking particles or resonances of different spin. A given potential may give rise to several trajectories, while a proper treatment of the theory shows that the J-values of successive

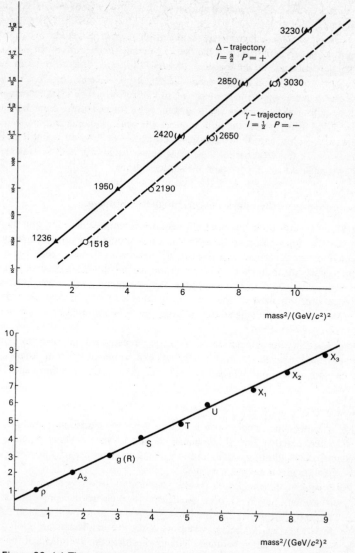

Figure 96 (a) The mass-squared values as a function of spin J for certain of the strangeness-zero baryon resonances. For the bracketed points the spin is unknown or uncertain. (b) The mass squared for the peaks found in the CERN missing-mass experiments, plotted as a function of peak number, probably equivalent to J, although only for ρ, A_2 and probably g, is the spin known

resonances on a given trajectory should jump by intervals of two rather than in single units.

In view of the rather vague general explanation given in the preceding paragraphs, without a proper explanation of the theory the real justification of the Regge idea for the student must be its apparent success. The trajectories for certain baryon resonances are shown in Figure 96(a). The peaks discovered in the CERN missing-mass spectrometer experiment are shown in Figure 96(b). The spins corresponding to these supposed resonances are in general not known, but the remarkable regularity of the straight-line relation between the mass squared and peak number (presumably equivalent to J) suggests that here too we have an example of Regge trajectories. In this case particles occur at unit 'J' intervals, and we may suppose that two adjacent trajectories are in fact degenerate.

10.10 Developments of the quark model: magnetic moments

The SU(3) classification has been enlarged to encompass also the spin quantum number. The appropriate group is SU(6), which should be a good symmetry if the interactions are invariant with respect to rotations of the spin and if spin–orbit couplings are not important. This limitation means that SU(6) can be valid only in the non-relativistic region, unlike the SU(3) symmetry which was independent of space–time. In view of this limitation the successes achieved by SU(6) are surprising, since one might expect that the tightly bound quarks would be highly relativistic.

In considering the calculation of magnetic moments we shall need to construct the appropriate states for proton and neutron from the spinning quarks. With certain simple assumptions SU(3) itself can give certain relations such as for instance

$$\mu(\Sigma^+) = \mu(p) \qquad \text{(measured values } 2{\cdot}5 \pm 0{\cdot}5 \text{ and } 2{\cdot}79 \text{ respectively).}$$

(This is immediately obvious if it is assumed that the electromagnetic interaction is a U-spin scalar.) The SU(6) quark model, however, while including these relations, yields additional ones and in particular relates the very precisely known proton and neutron magnetic moments.

We have seen that the proton is constructed from the quark combination (ppn). We now use the Clebsch–Gordan coefficients to construct the spin-$\frac{1}{2}$ quark states. We shall take $j_z = +\frac{1}{2}$ so that the $(\frac{1}{2}, +\frac{1}{2})$ physical proton will be built from the three states (A), (B), (C) below, where we make the simplest assumption that the quarks are all in relative s-states. (This implies that the overall wave function is symmetric whereas the generalized Pauli principle (3.7) would require (Fermion) spin-$\frac{1}{2}$ quarks to have an antisymmetric wave function. It is not clear why the quarks should differ from all other particles in this respect.)

	'p'	'p'	'n'
(A)	$(\frac{1}{2}, +\frac{1}{2})$	$(\frac{1}{2}, +\frac{1}{2})$	$(\frac{1}{2}, -\frac{1}{2})$
(B)	$(\frac{1}{2}, +\frac{1}{2})$	$(\frac{1}{2}, -\frac{1}{2})$	$(\frac{1}{2}, +\frac{1}{2})$
(C)	$(\frac{1}{2}, -\frac{1}{2})$	$(\frac{1}{2}, +\frac{1}{2})$	$(\frac{1}{2}, +\frac{1}{2})$.

States in which the proton and neutron quarks are arranged in a different order, e.g. pnp, are all the same, the only point of consequence for this problem being the spin orientations.

Grouping now the two proton quarks we see that

$$(A) \equiv (1, +1)(\tfrac{1}{2}, -\tfrac{1}{2}),$$

(B) and (C) $\equiv (1, 0)(\tfrac{1}{2}, +\tfrac{1}{2})$.

We can write

$$(1, 0)(\tfrac{1}{2}, +\tfrac{1}{2}) = \tfrac{1}{2}(B) + \tfrac{1}{2}(C).$$

Thus referring to the table of Clebsch–Gordan coefficients we have

$$\begin{aligned}
p_{\frac{1}{2}} &= \sqrt{\tfrac{2}{3}}(A) - \sqrt{\tfrac{1}{3}}\sqrt{\tfrac{1}{2}}[(B) + (C)] \\
&= \tfrac{1}{6}(2A - B - C) \\
&= \tfrac{1}{6}(2p\uparrow p\uparrow n\downarrow - p\uparrow p\downarrow n\uparrow - p\downarrow p\uparrow n\uparrow),
\end{aligned} \qquad \textbf{10.11}$$

where the arrows indicate the spin directions.

Similarly, the physical neutron may be written in terms of spinning quarks as

$$n_{\frac{1}{2}} = \tfrac{1}{6}(2n\uparrow n\uparrow p\downarrow - n\uparrow n\downarrow p\uparrow - n\downarrow n\uparrow p\uparrow). \qquad \textbf{10.12}$$

If we write a basic 'quark magneton' in the usual way as

$$\mu_q = \frac{Q\hbar}{2m_q c},$$

where Q is the charge and m_q the mass of the quark, then the proton, neutron and Λ-quarks will have magnetic moments expressed in terms of μ_q and the z-component of the spin σ_3 as

$$\mu_p = \tfrac{2}{3}\sigma_{3,p}\mu_q,$$
$$\mu_n = -\tfrac{1}{3}\sigma_{3,n}\mu_q,$$
$$\mu_\Lambda = -\tfrac{1}{3}\sigma_{3,\Lambda}\mu_q.$$

We now calculate the proton and neutron magnetic moments as the sums of the moments arising from the quark states of **10.11** and **10.12**,

$$\begin{aligned}
\mu_p &= \tfrac{1}{6}[4(\tfrac{2}{3} + \tfrac{2}{3} + \tfrac{1}{3}) + (\tfrac{2}{3} - \tfrac{2}{3} - \tfrac{1}{3}) + (-\tfrac{2}{3} + \tfrac{2}{3} - \tfrac{1}{3})]\mu_q \\
&= \mu_q \\
\mu_n &= -\tfrac{2}{3}\mu_q,
\end{aligned}$$

so that $\mu_n/\mu_p = -\tfrac{2}{3}$. The experimental value is $-1\cdot913/2\cdot792 = -0\cdot68$. The agreement represents a further striking success for the quark model.

10.11   The search for quarks

In the preceding sections we have discussed the successes of the quark model in the classification of resonances and in the calculation of some magnetic moments.

These are the most striking results although others, such as further magnetic-moment calculations, electromagnetic mass differences and interaction characteristics based on a model of interactions between quark pairs, also lend strong support to the model.

There are physicists for whom the existence or not of actual quarks is irrelevant and who would be happy to frame the theory in terms of quarks as purely mathematical constructs. The history of physics suggests otherwise and it is the author's view that if the quark model is successful then real quarks will in due course be found.

The most distinctive readily observable property of the quarks is their fractional charge. The ionization produced by the $\frac{1}{3}e$ and $\frac{2}{3}e$ quarks will be $\frac{1}{9}$ and $\frac{4}{9}$ that of singly charged particles travelling at the same speed. *A priori* it is not clear what may be the quark mass, since the interquark forces are certainly very strong and the $q\bar{q}$ and qq binding energies may be very great, rendering possible quite low mass $q\bar{q}$ meson resonances formed from very high mass quarks.

If the quarks are indeed very heavy then it may be difficult or impossible to create $q\bar{q}$ pairs using existing accelerators, since even if the threshold for $q\bar{q}$ production is accessible the cross-section, as suggested by any statistical type theory, is likely to be small.

Two general methods of quark search have been pursued:

(a) the search for less than minimum ionizing particles in bubble chambers, cloud chambers, spark chambers and counters, either using accelerators or very high energy cosmic rays

(b) the search for $\frac{1}{3}e$ and $\frac{2}{3}e$ charge objects in material. This latter method assumes that since the quarks have fractional baryon number and charge they cannot annihilate in interactions with ordinary matter.

We describe below some few examples of the many experiments of each kind of search.

10.11.1   *Particles of less than minimum ionization*

To illustrate the problems involved, we first discuss briefly a bubble-chamber search carried out at the CERN proton synchrotron (e.g. Blum *et al.*, 1964). In this experiment photographs in a liquid-hydrogen bubble chamber containing 1.5×10^5 high-momentum tracks from a target in the CERN PS were searched for low-density tracks. Nineteen such tracks were observed. However, the sensitivity of a bubble chamber varies throughout its pressure cycle, so that early or late tracks, which do not fall within the region of maximum sensitivity, will appear with low density. During the study in question the beam particles passed through scintillation counters and were recorded as a function of time on a display oscilloscope. In each of the nineteen pictures which showed a low-density track the oscilloscope showed an 'early' beam particle arriving between 1 and 2 ms before maximum sensitivity.

A somewhat different technique has been used in an experiment at the 70 GeV Serpukhov accelerator in the USSR. A beam channel was set up, for a momentum of 80 GeV/c, for singly charged particles in excess of the maximum momentum of singly charged secondaries from the accelerator target. For particles of charge $\frac{1}{3}e$ and $\frac{2}{3}e$ the momenta accepted are 26·7 and 53·4 GeV/c respectively, so that the background in the quark search is severely reduced (a very few particles scattered on the collimator walls or passing through shielding still reach the detectors).

Ionization losses were measured in a series of scintillation counters while the particle time-of-flight was also measured. A Cerenkov counter was used to suppress the background of light particles while a magnetic spectrometer with wide-gap spark chambers allowed the particle momentum to be measured to ± 1 per cent.

In this experiment, no event corresponding to a particle with fractional charge $\frac{1}{3}e$ or $\frac{2}{3}e$ was found. From this data the total cross-section for quark production with charge $-\frac{1}{3}e$, and mass 4·5–5 GeV/c^2, is found to be less than 3×10^{-37} mm^2 for 70 GeV protons on aluminium (Antipov et al., 1969).

Many experiments have searched for quarks in the cosmic radiation, where one may have across to particles from very high energy interactions but where the flux is small compared with that available from accelerators. With one exception, all these experiments have given null results. The one experiment in which apparently below-minimum-ionization tracks have been found is a study of tracks occurring in the cores of high-energy air showers due to cosmic-ray particles of extremely high energy interacting near the top of the atmosphere (McCusker and Cairns, 1969). An expansion cloud chamber was triggered by a counter array which detected the shower. The cloud chamber was expanded 100 ms after the shower and the photograph taken 100 ms after expansion. During a year of operation four low-density tracks were observed, of which an example is shown in Figure 97. The age of the track is determined by measuring the diffusive width. This measurement establishes it as arriving at the same time as the more heavily ionizing particles. By comparison with nearby tracks passing above and below it, the light track is shown to be in a well-illuminated region of the chamber.

Droplet counts in the light and other nearby tracks yield a ratio of $0\cdot48\pm0\cdot05$. These counts are made only in regions where δ-rays (as indicated by the denser patches in the track) are absent. Several points concerning these events remain to be resolved:

(a) The relativistic rise in ionization density in the chamber gas may be as great as 50 per cent in which case, if the low-density track were at minimum while the higher-density tracks were very relativistic, the comparison of drop densities would be $\sim 110/(230/1\cdot5) = 110/153 = 0\cdot72\pm0\cdot08$, and might just possibly represent a statistical fluctuation.

(b) It is not known whether the statistical distribution in drop counts is Gaussian.

(c) The exclusion of the δ-ray regions of tracks is to some extent subjective.

Hopefully, further studies will resolve the above points and resolve whether or not such tracks are due to fractional-charge particles.

The energy of the primary particles responsible for the showers used to trigger

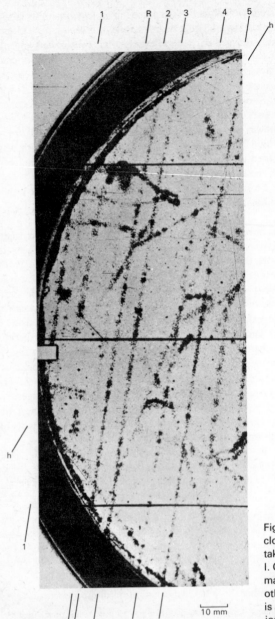

Figure 97 A Wilson cloud-chamber photograph taken by C. McCusker and I. Cairns (1969). The track marked R–R is parallel to the other tracks in the picture but is apparently of lower ionization

10 mm

the cloud chambers in the work of McCusker and Cairns, described above, was $\sim 3 \times 10^6$ GeV. *If* such energies represented the threshold for quark formation, the study of quarks would depend on cosmic rays for the foreseeable future.

10.11.2   *Fractional charges in matter*

If quarks have been bombarding the earth (or bombarding the meteorites in space) for a long period, then we might expect to find a small concentration of quarks in matter. Many samples of different material, ranging from meteorites to deep ocean sludge to dust from air filters, have been searched for fractional charges.

Several methods have been used such as the study of optical spectra, the Millikan experiment and other magnetic and electrostatic techniques. One of the most elegant is the magnetic levitometer (e.g. Becci, Gallinaro and Morpurgo, 1965; Gallinaro and Morpurgo, 1966).

In this experiment a magnetic well is created, using a magnet having the form shown in Figure 98(a). The equipotential contours in the xOz and yOz planes are shown in 98(b). A small particle of a diamagnetic substance, such as graphite, is

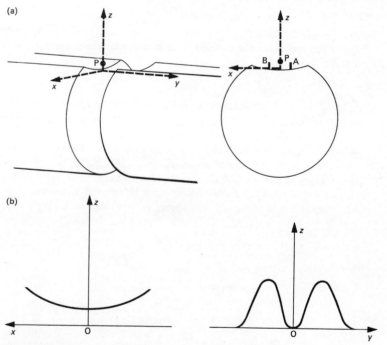

Figure 98 (a) The magnet in which the particle P is levitated and moved in the electrostatic field between the plates A and B (Becci, Gallinaro and Morpurgo, 1965). (b) The equipotential contours in the magnet of Figure 98(a)

inserted into the well between the two small plates A and B between which is an electric field of 10^5Vm^{-1}, which can be used to move the particle. The component of the gravitational force normal to the equipotential is balanced by the magnetic force. For this reason much larger sample particles can be used than in a Millikan experiment, where the whole mass must be supported by the electric field. In the latest form of the experiment, grains of 5×10^{-7} g can be handled which is about 50 000 times the mass of a typical Millikan droplet. The particle sits near the bottom of the almost circular equipotential and the tangential component of the gravitational force is balanced by the force due to the electric field. If the particle is displaced by a distance d from the equilibrium position then

$$d = \frac{qE}{mg} \times R,$$

where q is the particle charge, E the field, m the mass, g the acceleration due to gravity and R the radius of curvature of the equipotential; d is measured by a microscope and for a singly charged particle is typically 250μm. For a given particle we then normally have

$$d = n\delta,$$

where δ is the displacement for a single electron charge and n is the number of charges. This number can be altered by means of a radioactive source. If the particle contains a quark then

$$d = (n \pm \tfrac{1}{3})\delta.$$

Amplifications of the displacement may be achieved by applying an alternating field with a frequency near the natural frequency of the particle. The instrument in this form is sensitive to $\frac{1}{20}$ of the electronic charge.

A large number of particles have been examined in which no quarks have been found, setting a present limit of no quarks in 2×10^{18} nucleons.

In summary we can say that there is at present no *conclusive direct* evidence for the existence of quarks and their reality must be inferred in indirect ways. This presumably is due to the quark mass being greater than $\sim 5 \text{ GeV}/c^2$.

Appendix A
Relativistic kinematics
and phase space

A.1 Lorentz transformations

In elementary-particle physics we frequently have to deal with problems of relativistic particle kinematics. Most frequently the problems we meet are concerned with transformations between the laboratory system, where measurements are made, and the centre-of-momentum system (often misleadingly but conveniently called the centre-of-mass system), or c.m.s., to which the theoretical predictions apply directly.

Lorentz transformations are transformations which satisfy the basic postulate of special relativity theory that the velocity of light is the same for all observers, regardless of their state of relative motion. The particular underlying importance of the proper Lorentz transformations lies in the fact that they correspond to the transformations associated with the operators of quantum mechanics, so that for Lorentz transformations a true physical law will have the same form in systems moving relative to each other.

Consider two systems, (1) and (2), moving with uniform relative translational motion (Figure 99). In the classical case the positions of A in the two systems are clearly related by

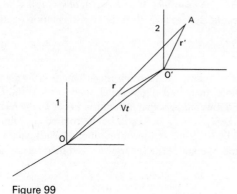

Figure 99

$$\mathbf{r} = \mathbf{r}' + \mathbf{V}t. \tag{A.1}$$

$\ddot{\mathbf{r}} = \ddot{\mathbf{r}}'$, so that Newton's laws hold in both systems. However the relativity postulate does not hold. Suppose a light pulse is emitted from O which after a time t has a spherical waveform on which lies A. Then differentiating **A.1** with regard to time we have

$$\mathbf{c} = \mathbf{c}' + \mathbf{V}.$$

For a satisfactory relativistic theory we proceed as follows: Suppose that O and O′ are coincident at time $t = 0$ and that at this instant a light pulse is emitted from O. In (1) the wavefront has the equation

$$r^2 = x^2 + y^2 + z^2 = c^2 t^2$$

or $\quad r^2 - c^2 t^2 = x^2 + y^2 + z^2 - c^2 t^2 = 0.$

If c is the same in both systems, then in (2) the wavefront has equation

$$r'^2 - c^2 t'^2 = x'^2 + y'^2 + z'^2 - c^2 t'^2 = 0.$$

Thus $\qquad\qquad r'^2 - c^2 t'^2 = r^2 - c^2 t^2$

or $\quad x'^2 + y'^2 + z'^2 - c^2 t'^2 = x^2 + y^2 + z^2 - c^2 t^2.$

We can write this as

$$\sum_{i=1}^{4} x_i^2 = \sum_{i=1}^{4} x_i'^2 = \text{constant}, \tag{A.2}$$

where we have written

$$x = x_1, \quad y = x_2, \quad z = x_3, \quad ict = x_4.$$

The x_i are the components of a *Lorentz-invariant four-vector*. We shall discuss other such four-vectors below.

Let us examine the *transformation operators*. If we have an observable Q in one system and the value in the second system is Q' then the transformation operator L is such that

$$Q' = LQ. \tag{A.3}$$

Since for invariant four-vectors we have the relationship **A.2** the transformations must be equivalent to rotations in a four-dimensional space with orthogonal axes along the x_i directions. First we make a spatial rotation to get V, the relative velocity along x_3. We now take x_1, x_2, x_3 and x_4 to be the rotated coordinates. x_1 and x_2 are then unaffected by the Lorentz transformation, which is equivalent to a rotation in the x_3, x_4 plane. We now write the transformed quantities x' as

$$x_j' = \sum_{i=1}^{4} a_{ji} x_i, \tag{A.4}$$

where the a_{ji} are linked by the orthogonality conditions

$$\sum_i a_{ji}a_{ki} = \delta_{jk}. \qquad \textbf{A.5}$$

Since x'_3 and x'_4 cannot depend on x_1 and x_2, L can be shown to be a 4×4 matrix having the form

$$L = \begin{bmatrix} 1 & 0 & 0 & 0 \\ 0 & 1 & 0 & 0 \\ 0 & 0 & a_{33} & a_{34} \\ 0 & 0 & a_{43} & a_{44} \end{bmatrix}. \qquad \textbf{A.6}$$

This transformation operator can be applied to any four-vector. If we take the particular case of the (x, y, z, ict) four-vector, we can write

$$x_3 = Vt = -i\beta x_4, \qquad \textbf{A.7}$$

where $\beta = V/c$.

Using equations **A.3–7** the student should show that

$$a_{33} = a_{44} = \frac{1}{\sqrt{(1-\beta^2)}} = \gamma,$$

$$a_{34} = i\beta\gamma \quad \text{and} \quad a_{43} = -i\beta\gamma.$$

One of the most useful of four-vectors is that formed by the energy and the three components of the momentum. If we write the total energy as ε, the momentum as p and the rest mass as m then we can write

$$\varepsilon^2 - p^2 c^2 = m^2 c^4.$$

We may choose to use units such that $c = 1$ and write this as

$$\varepsilon^2 - p^2 = m^2.$$

The rest mass is invariant, so that $(p_x, p_y, p_z, i\varepsilon) = (\mathbf{p}, i\varepsilon)$ is a Lorentz-invariant four-vector.

Writing

$$\mathbf{p} = m\mathbf{v}\gamma = m\boldsymbol{\beta}\gamma \quad \text{and} \quad \varepsilon = m\gamma,$$

where $\mathbf{v}$ is a particle velocity we have

$$\mathbf{v} = \mathbf{p}/\varepsilon.$$

Also for a system of particles with total energy E and total momentum $\mathbf{P}$ the velocity $\mathbf{V}$ is given by $\mathbf{P}/E$. Thus for two particles

$$\mathbf{V}_{\text{c.m.s.}} = \frac{\mathbf{P}}{E} = \frac{\mathbf{p}_1 + \mathbf{p}_2}{\varepsilon_1 + \varepsilon_2}.$$

Suppose we now have a particle moving with momentum p at an angle θ to the x_3 axis, which is the direction of motion of the centre of momentum. Then if V is the c.m.s. velocity we can apply the operator L to derive the following transformation equations, linking laboratory scalar momentum p, the angle θ and the energy ε to the c.m.s. quantities. We find

$$\varepsilon' = \gamma(\varepsilon - Vp\cos\theta),$$
$$p'\cos\theta' = \gamma(p\cos\theta - V\varepsilon),$$
$$p'\sin\theta' = p\sin\theta,$$
$$\varepsilon = \gamma(\varepsilon' + Vp'\cos\theta'),$$
$$p\cos\theta = \gamma(p'\cos\theta' + V\varepsilon').$$

$(\mathbf{V}.\mathbf{p} = Vp\cos\theta \quad \text{and} \quad \mathbf{V}.\mathbf{p}' = Vp'\cos\theta').$

The student should check these relations as an exercise.

A.2 The centre-of-mass system

In the c.m.s. the total momentum is zero; for a two-particle collision the particles approach each other with equal and opposite momenta. This system is unique, and calculations which depend on energy and momentum are considerably simplified by its use. We use the following notation:

E, ε: total energy of system or particle

T, t: kinetic energy of system or particle

$\mathbf{P}, \mathbf{p}$: momentum of system or particle

M, m: mass of system or particle

$\mathbf{V}, \mathbf{v}$: velocity of c.m.s., velocity of particle.

As before, we write $c = 1$ and use dashed quantities for the c.m.s. and undashed quantities for the laboratory system.
Then

$$\varepsilon^2 = p^2 + m^2, \qquad \varepsilon = t + m, \qquad p^2 = t^2 + 2mt.$$

Using the energy momentum four-vector invariance we have

$$E^2 - P^2 = E'^2 - P'^2.$$

But, by definition, $P' = 0$ so that E', the energy available in the c.m.s., is given by

$$E'^2 = E^2 - P^2.$$

For example, consider a collision between a particle $m_1, v_1, \varepsilon_1, \mathbf{p}_1$ in the laboratory and a stationary particle $m_2, v_2 = 0, \varepsilon_2 = m_2, \mathbf{p}_2 = 0$. Then

$$E'^2 = \left(\sum\varepsilon\right)^2 - \left(\sum\mathbf{p}\right)^2$$
$$= (\varepsilon_1 + m_2)^2 - p_1^2.$$

Note that in the system of units with $c = 1$ we express momenta in units of MeV/c

and masses in units of MeV/c^2. Suppose we wish to determine the threshold momentum for the process

$$\pi^- p \rightarrow K^0 \Lambda^0.$$

The reaction will become possible when E' is just adequate to provide the masses of the secondary particles,

$$E' = m_\Lambda + m_K = 1\cdot613.$$

Writing $E'^2 = \varepsilon_1^2 - p_1^2 + m_2^2 + 2\varepsilon_1 m_2$

we have $\varepsilon_1 = \dfrac{1}{2m_2}(E'^2 - m_1^2 - m_2^2)$

$$= 0\cdot904 \text{ GeV}.$$

Thus $t_1 = 0\cdot764 \text{ GeV}, \quad p = 0\cdot89 \text{ GeV}/c.$

The following easily remembered formula for the threshold kinetic energy for a process may be derived as an exercise. If the total initial and final masses are M_i and M_f respectively, while the mass of the target particle is m_s, then

$$T_{\text{threshold}} = \frac{(M_f - M_i)(M_f + M_i)}{2m_s}.$$

The slow increase of E' with bombarding energy, for a stationary target, illustrates the advantage of clashing-beam experiments possible with storage rings, where E' is simply the sum of the energies of the clashing particles (if they are of equal mass).

A.3   Geometrical picture of transformations

A method of gaining a picture of the effect of the laboratory–c.m.s. transformation, as well as a simple way of deriving certain formulae, is afforded by the momentum ellipsoid proposed by Blaton (1950).

We first represent the non-relativistic situation in Figure 100.

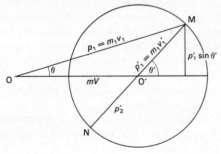

Figure 100

We draw a velocity-vector diagram where OO′ represents the velocity of the c.m.s. and OM the velocity of a particle in the laboratory O′M then represents the velocity in the c.m.s. Multiplying these vectors by m_1, the rest mass of the particle, OM now represents the laboratory momentum and O′M the c.m.s. momentum. In a two-particle process the momentum of the other particle must be given by O′N in the c.m.s. As θ and ϕ, the azimuthal angle, vary, M and N move on the surface of the sphere centred on O′ with radius p'.

We can readily generalize to the relativistic case. With V along the x-axis we have the relations

$$p'_x = \frac{1}{\gamma}(p_x - V\gamma\varepsilon'), \quad p'_y = p_y, \quad p'_z = p_z$$

so that $p'^2 = p'^2_x + p'^2_y + p'^2_z$

$$= p^2_z + p^2_y + \frac{1}{\gamma^2}(p_x - V\gamma\varepsilon')^2$$

or $\dfrac{p^2_z + p^2_y}{p'^2} + \dfrac{(p_x - V\gamma\varepsilon')^2}{p'^2\gamma^2} = 1,$

which is the equation of a prolate ellipsoid of revolution with axis along the x-axis and its centre displaced a distance $V\gamma\varepsilon'$ from the origin of the p_x vector.. The section in the xy plane is shown in Figure 101. The size and shape of the ellipsoid

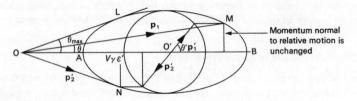

Figure 101

are determined by the masses and energies available in the reaction. When two particles emerge one of which has c.m.s. momentum $\mathbf{p}'_1$, the other must have momentum $\mathbf{p}'_2$. These can be seen at once to transform to the laboratory momenta $\mathbf{p}_1 = $ OM and $\mathbf{p}_2 = $ ON. It is also clear that the maximum and minimum laboratory momenta are OA and OB, and the maximum angle of emission of a particle in the laboratory corresponds to the tangent OL and is thus LÔA.

The student can show that the eccentricity of the ellipse is V and that the ratio of the semi-major to the semi-minor axes $a/b = \gamma$.

It is useful to distinguish two different geometrical configurations corresponding to different physical situations. If we write $\alpha = V\gamma\varepsilon'$, then if $\alpha > a$, O is outside the ellipse as in Figure 101. Since $a = p'\gamma$,

$\alpha > a$ corresponds to $V\varepsilon' > p'$,

i.e. $V > v'$,

the velocity of the centre-of-momentum is greater than the velocity of the particle in the c.m.s. In this case all particles go forward in the laboratory and there exists a maximum angle of emission. If $\alpha < a$ then O is inside the ellipsoid, the velocity of the c.m.s. is less than the particle velocity in the c.m.s., and θ can vary from 0 to π.

From the geometry of the ellipse (or by differentiating the expression for tan θ) it can be shown that

$$\sin \theta_{max} = \sqrt{\left(\frac{1-V^2}{\gamma_n^2 - V^2}\right)} = \frac{b^2}{fm} , \qquad\qquad\text{A.8}$$

where we have written γ_n for V/v'_n, where v'_n is the c.m.s. velocity, m the mass of the particle and f is the focal distance of the ellipse.

A.4   Decay length

As a final application of the momentum ellipsoid we may use it to visualize the variation of the decay length with angle for unstable particles emitted from a reaction and decaying in flight.

We define the *decay length l* in the laboratory system as the distance travelled in one mean life. Then

$$l_{lab} = v_{lab}\tau.$$

As is well known, a moving clock appears to the 'stationary' observer to run slow. If τ_0 is the mean life of the particle at rest then

$$\tau = \gamma\tau_0$$

$$\text{and} \quad l = v\gamma\tau_0 = \frac{p}{m}\tau_0 = p\frac{\tau_0}{m},$$

where p is the laboratory momentum. Thus

$$l \propto p$$

and if we look on the momentum vector diagram as a space decay diagram then the ellipsoid is the locus of the decay points of all particles which live exactly one mean life. For instance, consider the decay in flight of Σ-hyperons produced in the reactions

$$K^-p \begin{array}{c} \nearrow \Sigma^+\pi^- \\ \searrow \Sigma^-\pi^+, \end{array}$$

by K^- mesons of a given momentum. A simple calculation gives us p', the Σ-momentum in the c.m.s. We can now draw the ellipse with semi-minor axis scaled to equal

$$v'\tau_0 = \frac{p'}{m}\tau_0.$$

The ellipse will then represent the locus of decay points for Σs living for one mean life.

A.5 Limiting kinematic relationships for many-particle systems

The relationships discussed above are general for all laboratory–c.m.s. transformations. For reactions from which only two particles emerge they have a direct application since the c.m.s. momenta of the secondary particles are equal and opposite. Thus in a given reaction between two particles of fixed energy there is a *unique relationship* between the angles and momenta of the emerging particles given by the foregoing equations or the ellipse diagrams.

If three or more particles emerge there are no longer unique relationships of this kind but only limiting relationships. As an example we calculate the *maximum momentum* of the $(j+1)$th particle in the c.m.s. for a reaction from which $(j+1)$ particles emerge. The basis of this calculation is that any particle will attain its maximum momentum for a fixed total energy E' of all particles in the c.m.s., when all the other particles go off *together* in the opposite direction.

Thus $\quad v'_1 = v'_2 = v'_3 = \ldots = v'_j$

so that $\quad \dfrac{p'_1}{\varepsilon'_1} = \dfrac{p'_2}{\varepsilon'_2} = \ldots = \dfrac{p'_j}{\varepsilon'_j}$

and $\quad \dfrac{p'_1}{m_1} = \dfrac{p'_2}{m_2} = \ldots = \dfrac{p'_j}{m_j}.$

We can thus write all the 1 to j momenta in terms of p'_1 and the masses

$$p'_j = \frac{m_j}{m_1} p'_1, \text{ etc.}$$

Then for maximum $p'_{j+1} = p'_{max}$

$$p'_{max} = p'_1 + p'_2 \ldots + p'_j$$

$$= p'_1 \left(1 + \sum_{i=2}^{j} \frac{m_i}{m_1}\right). \tag{A.9}$$

Also $\quad E' = \varepsilon'_1 + \varepsilon'_2 + \ldots + \varepsilon'_j + \varepsilon'_{j+1}$

$$= (p'^2_1 + m^2_1)^{\frac{1}{2}} + \sum_{i=2}^{j} \left(\frac{m_i^2}{m_1^2} p'^2_1 + m_i^2\right)^{\frac{1}{2}} + \left[\left(1 + \sum_{i=2}^{j} \frac{m_i}{m_1}\right)^2 p'^2_1 + m^2_{j+1}\right]^{\frac{1}{2}}. \tag{A.10}$$

We can eliminate p'_1 from equations **A.9** and **A.10** to solve for p'_{max} as

$$p'^2_{max} = \frac{\left[E'^2 - \left(m + \sum_{i=1}^{j} m_i\right)^2\right]\left[E'^2 - \left(m - \sum_{i=1}^{j} m_i\right)^2\right]}{4E'^2}$$

or $\quad \varepsilon'_{max} = \dfrac{E'^2 + m^2 - \left(\sum_{i=1}^{j} m_i\right)^2}{2E'},$

where we have written $m = m_{j+1}$. In this limiting situation we have effectively a two-body process so that the maximum of all $\sin \theta_{max}$ is given by substituting the velocity corresponding to the above ε'_{max} into equation **A.8**.

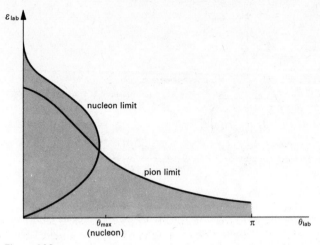

Figure 102

As an example we show in Figure 102 the accessible kinematic regions for particles from the reaction $pp \rightarrow pn\pi^+$, where the incident energy has been chosen such that for the nucleon $V > v_{n(max)}$. Thus at any angle the limiting value of ε_n is double valued as one would expect from the ellipsoid. For the pion $V < v_{max}$, ε_π is thus single-valued and there is no θ_{max}.

A.6 Processes with three particles in the final state: the Dalitz plot

We have seen that, for three or more particles in the final state, only limiting kinematic relationships may be given. A particularly useful way of treating the three-particle case was proposed by Dalitz. The particular merit of this kind of plot arises from the uniform distribution of events within it if the reaction proceeds according to the available density of states. We shall return to this topic in the next section. Here we deal only with the plot limits.

We first consider the case in which all three outgoing particles have the same mass, such as for instance the τ-decay of the K-meson into three pions. If the total energy in the c.m.s. is $E' (= m_\tau + T_\tau)$ then

$$E' = \varepsilon_1' + \varepsilon_2' + \varepsilon_3'$$

and $E' - \sum_{i=1}^{3} m_i = T' = t_1' + t_2' + t_3'.$

T' is a constant and so the three-particle kinetic energies in the c.m.s. can be represented as perpendicular distances from a point within a triangle to the three sides. With the x- and y-axes drawn as in Figure 103 we see that $y = t_1'$ and

$$x = \frac{1}{\sqrt{3}} (t_3' - t_2').$$

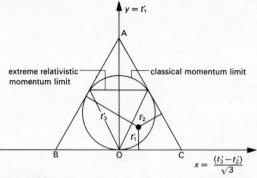

Figure 103

The conservation of energy required can then be satisfied for all points within the triangle. However the conservation of momentum imposes a tighter limit on the allowed region. In the classical case this limit is the inscribed circle, as can be seen by noting that the momentum limit is set by the condition

$$\mathbf{p}_1 + \mathbf{p}_2 + \mathbf{p}_3 = 0, \qquad\qquad\qquad \textbf{A.11}$$

so that in the classical case

$$(m\mathbf{v}_1 + m\mathbf{v}_2)^2 = m^2\mathbf{v}_3^2,$$

from which

$$t_1^2 + t_2^2 + t_3^2 - 2(t_1 t_2 + t_2 t_3 + t_3 t_1) = 0,$$

which is the inscribed circle

$$x^2 + \left(y - \frac{t_1 + t_2 + t_3}{3}\right)^2 = \left(\frac{t_1 + t_2 + t_3}{3}\right)^2.$$

In the extreme relativistic case we get a triangle, as shown in the figure, while in the usual intermediate situation we have a shape between these extremes, the equation of which may again be calculated by applying equation **A.11**.

Before leaving the symmetrical case we note some other features of the plot. If in the c.m.s. we group two particles which then have relative momentum $\mathbf{q}$, and with respect to which the third particle has momentum $\mathbf{p}$, we have the configuration

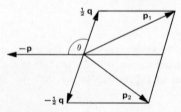

Figure 104

shown in Figure 104. $\mathbf{p}_1$, $\mathbf{p}_2$ and $-\mathbf{p}$ are then the momenta in the overall c.m.s. The student may check that

$$p^2 \propto PN,$$
$$q^2 \propto PQ,$$
$$\cos \theta = \frac{GP}{GH},$$

A.12

where P, N, Q, G, H are as shown in Figure 105. The dependences of **A.12** mean

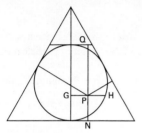

Figure 105

that if, for instance, the cross-section for a process is a function of $\cos \theta$, then this will be reflected as a variation in the density of the population across the circle.

If the masses of the three particles involved are not equal, the Dalitz plot is no longer symmetrical and there is then little merit in plotting

$$\frac{1}{\sqrt{3}} (t'_2 - t'_3)$$

along the x-axis rather than simply t'_2 say. Figure 106 shows the Dalitz plot for a

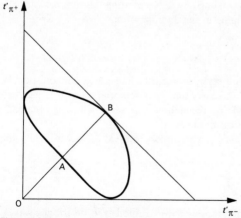

Figure 106

process such as $K^-p \rightarrow \Lambda\pi^+\pi^-$. In this case we have plotted t'_{π^+} against t'_{π^-} so that the plot is symmetrical about the 45° line. As before the boundary corresponds to collinear particles. At A, $t'_{\pi^-} = t'_{\pi^+}$ and t'_Λ is a maximum; at B, $t'_\Lambda = 0$. Along the tangent at B,

$$t'_{\pi^+} + t'_{\pi^-} = \text{constant} = E' - t'_\Lambda,$$

so t'_Λ is constant along this and parallel lines, increasing from zero at B to a maximum at A.

A.7 Phase space

Having examined the kinematic limits we now turn to the distribution within the limits, that is to the probability of the occurrence of a given set of particle momenta within the allowed range of momenta.

We start from the formula for the transition rate from the initial to the final state

$$T = \frac{2\pi}{\hbar} |M|^2 \frac{dN}{dE}. \qquad \text{A.13}$$

$|M|^2$ is the square of the modulus of the matrix element linking the states. dN/dE is the density of final states or 'phase-space' factor. If M is not a function of the final-state momenta, then such a dependence will be contained only in the phase-space factor.

It is convenient to work in a phase space where the coordinates represent the momentum components of a given particle. If there are n particles in the final state, then it can be represented by a vector in a $3n$-dimensional phase space. Imposition of energy conservation limits this vector, so that all possible states are seen to lie within the volume swept out by it in the $3n$-dimensional space. The number of states is then obtained by dividing this volume by the volume per state, and the density of states is obtained by differentiating this number with respect to E. Thus if $Q(E)$ is the volume in the $3n$-dimensional space we write

$$Q(E) = \prod_{i=1}^{n} \int d\mathbf{p}_i,$$

where there are implicit limits set by momentum and energy conservation. We shall make these limits explicit, and carry out the differentiation with respect to energy, below, but shall first divide by the volume per state, which is obtained by the standard method of statistical mechanics.

The number of plane waves in a frequency interval $v - v + \delta v$ within a volume V is given by

$$\delta N(v) = N(v)\,\delta v = 4\pi V \frac{v^2\,\delta v}{c^3}.$$

In terms of momentum this translates into a number of de Broglie waves

$$N(p)\,\delta p = \frac{4\pi V}{h^3}\,p^2\,\delta p = \frac{V}{(2\pi\hbar)^3}\,4\pi p^2\,\delta p.$$

These are contained in a phase-space element which is a shell of thickness δp, radius p and volume $4\pi p^2\,\delta p$. Each possible state corresponds to a discrete point in phase space, so that the discrete quantized volume occupied by each is $(2\pi\hbar)^3/V$. For the n-particle case, the unit cell volume is $(2\pi\hbar/V^{\frac{1}{3}})^{3n}$, and the number of states is

$$N = \left(\frac{V^{\frac{1}{3}}}{2\pi\hbar}\right)^{3n} Q(E).$$

We may impose the momentum and energy conservation explicitly by considering first one integral

$$G(E) = \int d^3\mathbf{p}_1\,\theta(x),$$

where $\theta(x)$ is a step function inserted to limit the states to the allowed region, and is illustrated in Figure 107. Only at A is $d\theta/dx$ different from zero, and at A

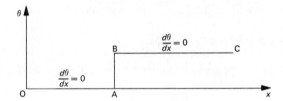

Figure 107

$$\frac{d\theta}{dx} = \delta(x_A).$$

Thus $\dfrac{dG(E)}{dE} = \displaystyle\int d^3\mathbf{p}_1\,\frac{d\theta(E'-\varepsilon_1)}{dE} = \int d^3\mathbf{p}_1\,\delta(E'-\varepsilon_1).$

We can ensure momentum conservation by including another δ-function. Returning also to the multiparticle equation we have

$$\frac{dN}{dE} = \left(\frac{V^{\frac{1}{3}}}{2\pi\hbar}\right)^{3n} \prod_{i=1}^{n} \int d^3\mathbf{p}_i\,\delta\left(E' - \sum_{i=1}^{n}\varepsilon_i\right)\delta^3\left(\sum_{i=1}^{n}\mathbf{p}_i\right). \qquad \textbf{A.14}$$

The expression for the density of states which follows from **A.14** is not itself Lorentz invariant, which means that, since the transition rate *is* Lorentz invariant, the matrix element itself must still contain kinematic factors.

It is easy to see qualitatively how the result must be modified to render the phase space or density of states factor invariant and thus extract such kinematic factors from the matrix element. The normal probability density is expressed in quantum mechanics as the square of the modulus of the wave function for each particle $|\psi|^2$, with normalization $\int |\psi|^2\, dx\, dy\, dz = 1$. Due to the relativistic contraction of a moving system, this density increases by a factor γ for a moving system so that the density is not invariant. However since the total energy ε also varies as γ, we can write a density $|\sqrt{\varepsilon}\psi|^2$ which is invariant. In fact it is convenient to use $|\sqrt{(2\varepsilon)}\psi|^2$. Then, introducing $\sqrt{(2\varepsilon_i)}$ to the matrix element for each particle, we have a Lorentz-invariant element M' and we write instead of **A.13** and **A.14**

$$T = \frac{2\pi}{\hbar}|M'|^2 \left(\frac{V^{\frac{4}{3}}}{2\pi\hbar}\right)^{3n} \prod_{i=1}^{n} \int \frac{d^3\mathbf{p}_i}{2\varepsilon_i}\, \delta\left(E' - \sum_{i=1}^{n} \varepsilon_i\right) \delta^3\left(\sum_{i=1}^{n} \mathbf{p}_i\right). \qquad \textbf{A.15}$$

We may also approach this equation by working directly in terms of the energy momentum four-vectors $\tilde{p}_i$ and writing the density of states factor as

$$\frac{dN}{dE} \propto \prod_{i=1}^{n-1} \int d^4\tilde{p}_i\, \delta(p_i^2 - m_i^2)$$

$$= \prod_{i=1}^{n-1} \int d^3\mathbf{p}_i\, d\varepsilon_i\, \delta(\varepsilon_i^2 - p_i^2 - m_i^2)$$

$$= \prod_{i=1}^{n-1} \int d^3\mathbf{p}_i\, \delta(\varepsilon_i^2 - p_i^2 - m_i^2)\, d(\varepsilon_i^2 - p_i^2 - m_i^2)\, \frac{d\varepsilon_i}{d(\varepsilon_i^2 - p_i^2 - m_i^2)}$$

$$= \prod_{i=1}^{n-1} \int \frac{d^3\mathbf{p}_i}{2\varepsilon_i}\, d(\varepsilon_i^2 - p_i^2 - m_i^2)\, \delta(\varepsilon_i^2 - p_i^2 - m_i^2).$$

But the integration over $(\varepsilon_i^2 - p_i^2 - m_i^2)$ is cancelled by the δ-function so that the expression becomes

$$\prod_{i=1}^{n-1} \int \frac{d^3\mathbf{p}_i}{2\varepsilon_i}\, \delta(\varepsilon_i^2 - p_i^2 - m_i^2),$$

which is directly related to equation **A.15**.

We now specialize to the three-particle case and integrate over the uninteresting variables.

$$\frac{dN}{dE} = \frac{V^3}{(2\pi\hbar)^9} \int \frac{d\mathbf{p}_1}{2\varepsilon_1} \int \frac{d\mathbf{p}_2}{2\varepsilon_2} \int \frac{d\mathbf{p}_3}{2\varepsilon_3}\, \delta\left(\sum_{i=1}^{3} \mathbf{p}_i\right) \delta\left(E' - \sum_{i=1}^{3} \varepsilon_i\right).$$

The momenta are related as in Figure 108 and

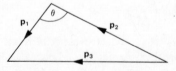

Figure 108

$$\delta\mathbf{p} = \delta p + p\delta\theta + p\sin\theta\,\delta\phi$$

so that $\displaystyle\int d\mathbf{p} = \int dp \int p\,d\theta \int p\sin\theta\,d\phi$

$$= \int p^2\,dp \int d\Omega.$$

Thus

$$\iint \frac{d\mathbf{p}_1}{\varepsilon_1}\frac{d\mathbf{p}_2}{\varepsilon_2} = \int \frac{4\pi p_1^2}{\varepsilon_1}\,dp_1 \int p_2^2\frac{dp_2}{\varepsilon_2} \int d\phi\,d(\cos\theta)$$

$$= \int \frac{4\pi p_1^2}{\varepsilon_1}\,dp_1 \int 2\pi\frac{p_2^2}{\varepsilon_2}\,dp_2 \int d(\cos\theta),$$

while the integration over $\mathbf{p}_3$ is now redundant since it is fixed.

To express the integration over $\cos\theta$ in terms of the other quantities we write

$$\varepsilon_3^2 = p_3^2 + m_3^2$$

$$= m_3^2 + p_1^2 + p_2^2 + 2p_1 p_2\cos\theta.$$

Then, taking differentials with respect to $\cos\theta$ and holding p_1 and p_2 fixed, we have

$$2p_1 p_2\,\delta(\cos\theta) = 2\varepsilon_3\,\delta\varepsilon_3.$$

Using $p\,\delta p = \varepsilon\,\delta\varepsilon$ (since $\varepsilon^2 = p^2 + m^2$), we have

$$\frac{dN}{dE} = \frac{V^3}{(2\pi\hbar)^9}\frac{8\pi^2}{8}\int d\varepsilon_1 \int d\varepsilon_2 \int d\varepsilon_3\,\delta\left(E' - \sum_{i=1}^{3}\varepsilon_i\right)$$

and finally dropping the redundant integration

$$T = \frac{2\pi}{\hbar}\,|M'|^2\,\frac{V^3}{(2\pi\hbar)^9}\,\pi^2 \iint d\varepsilon_1\,d\varepsilon_2,$$

i.e. $\displaystyle T \propto \iint d\varepsilon_1\,d\varepsilon_2$

$$= \iint dt_1\,dt_2.$$

But $\delta t_1\,\delta t_2$ is the element of area in the Dalitz plot so that:

If a process is governed solely by the density of states factor the density of the population in the Dalitz plot will be uniform. Any deviation from uniformity is indicative of a special interaction depending on the relative momenta of the particles, such as for instance a strong force binding two particles together.

It is easy to show that in the symmetrical Dalitz plot for three particles of equal mass, where $x \propto (t_3 - t_2)$ and $y \propto t_1$, then $\delta y \propto \delta t_1$ and $\delta x \propto \delta t_2$. The Dalitz plot is frequently made in terms of the effective-mass squared of particle pairs. Such a plot still has the property that the elementary area $\delta m_{ij}^2\,\delta m_{jk}^2$ is proportional to the element of phase space, since

$$m_{ij}^2 = (\varepsilon_i + \varepsilon_j)^2 - (\mathbf{p}_i + \mathbf{p}_j)^2$$
$$= (E - \varepsilon_k)^2 - p_k^2,$$
$$\delta m_{ij}^2 = -2(E - \varepsilon_k)\delta\varepsilon_k - 2p_k\delta p_k$$
$$= -2E\,\delta\varepsilon_k.$$

So that $\quad \delta m_{ij}^2\, \delta m_{jk}^2 \propto \delta\varepsilon_k\, \delta\varepsilon_i$

and thus the effective-mass squared Dalitz plot will also be uniformly populated if the reaction is regulated by phase space.

A number of examples of the Dalitz plot are shown in Chapter 9.

Appendix B
Clebsch–Gordan coefficients and particle properties

B.1 Clebsch–Gordan coefficients

$$\psi(J, M) = \sum_{m_1, m_2} C(J, M, j_1, m_1, j_2, m_2)\, \phi(j_1, m_1)\, \chi(j_2, m_2)$$

For each pair of values of j_1 and j_2 the tables give:

		J	J	...
		M	M	...
m_1	m_2
m_1	m_2
.	.
.	.

In all cases the *squares of the coefficients are given*.

(a) $j_1 = \frac{1}{2}, \quad j_2 = \frac{1}{2}$

		1	1	0	1
		+1	0	0	−1
$+\frac{1}{2}$	$+\frac{1}{2}$	1			
$+\frac{1}{2}$	$-\frac{1}{2}$		$\frac{1}{2}$	$\frac{1}{2}$	
$-\frac{1}{2}$	$+\frac{1}{2}$		$\frac{1}{2}$	$-\frac{1}{2}$	
$-\frac{1}{2}$	$-\frac{1}{2}$				1

(b) $j_1 = 1, \quad j_2 = \frac{1}{2}$

		$\frac{3}{2}$	$\frac{3}{2}$	$\frac{1}{2}$	$\frac{3}{2}$	$\frac{1}{2}$	$\frac{3}{2}$
		$+\frac{3}{2}$	$+\frac{1}{2}$	$+\frac{1}{2}$	$-\frac{1}{2}$	$-\frac{1}{2}$	$-\frac{3}{2}$
$+1$	$+\frac{1}{2}$	1					
$+1$	$-\frac{1}{2}$		$\frac{1}{3}$	$\frac{2}{3}$			
0	$+\frac{1}{2}$		$\frac{2}{3}$	$-\frac{1}{3}$			
0	$-\frac{1}{2}$				$\frac{2}{3}$	$\frac{1}{3}$	
-1	$+\frac{1}{2}$				$\frac{1}{3}$	$-\frac{2}{3}$	
-1	$-\frac{1}{2}$						1

(c) $j_1 = 1$, $j_2 = 1$

		2	2	1	2	1	0	2	1	2
		+2	+1	+1	0	0	0	-1	-1	-2
+1	+1	1								
+1	0		$\frac{1}{2}$	$\frac{1}{2}$						
0	+1		$\frac{1}{2}$	$-\frac{1}{2}$						
+1	-1				$\frac{1}{6}$	$\frac{1}{2}$	$\frac{1}{3}$			
0	0				$\frac{2}{3}$	0	$-\frac{1}{3}$			
-1	+1				$\frac{1}{6}$	$-\frac{1}{2}$	$\frac{1}{3}$			
0	-1							$\frac{1}{2}$	$\frac{1}{2}$	
-1	0							$\frac{1}{2}$	$-\frac{1}{2}$	
-1	-1									1

(d) $j_1 = \frac{3}{2}$, $j_2 = \frac{1}{2}$

		2	2	1	2	1	2	1	2
		+2	+1	+1	0	0	-1	-1	-2
$+\frac{3}{2}$	$+\frac{1}{2}$	1							
$+\frac{3}{2}$	$-\frac{1}{2}$		$\frac{1}{4}$	$\frac{3}{4}$					
$+\frac{1}{2}$	$+\frac{1}{2}$		$\frac{3}{4}$	$-\frac{1}{4}$					
$+\frac{1}{2}$	$-\frac{1}{2}$				$\frac{1}{2}$	$\frac{1}{2}$			
$-\frac{1}{2}$	$+\frac{1}{2}$				$\frac{1}{2}$	$-\frac{1}{2}$			
$-\frac{1}{2}$	$-\frac{1}{2}$						$\frac{3}{4}$	$\frac{1}{4}$	
$-\frac{3}{2}$	$+\frac{1}{2}$						$\frac{1}{4}$	$-\frac{3}{4}$	
$-\frac{3}{2}$	$-\frac{1}{2}$								1

B.2 Properties of the stable particles

B.2.1 *Mesons. leptons, photon ($B = 0$)*

	J^P	S	I	Mass/ (MeV/c^2)	Mean life/s	Decays	Fraction /per cent	Q/MeV
γ	1^-	0	0	0	stable			
ν_e	$\frac{1}{2}$		0	0	stable			
ν_μ	$\frac{1}{2}$		0	0	stable			
$e^\pm$	$\frac{1}{2}$		0	0·51	stable			
$\mu^\pm$	$\frac{1}{2}$		0	105·6	$2{\cdot}20 \times 10^{-6}$	$e\bar{\nu}\nu$	100	105
$\pi^\pm$	0^-	0	1	139·6	$2{\cdot}60 \times 10^{-8}$	$\mu\nu$	100	34
π^0	0^-	0	1	135·0	$0{\cdot}89 \times 10^{-16}$	$\gamma\gamma$	98·8	135
						$\gamma e^+ e^-$	1·2	134

	J^P	S	I	Mass/ (MeV/c^2)	Mean life/s	Decays	Fraction /per cent	Q/MeV
$K^{\pm}$	0^-	± 1	$\frac{1}{2}$	493·8	$1\cdot2\times10^{-8}$	$\mu\nu$	63·5	388
						$\pi^{\pm}\pi^0$	20·8	219
						$\pi^{\pm}\pi^+\pi^-$	5·5	75
						$\pi^{\pm}\pi^0\pi^0$	1·7	84
						$\mu^{\pm}\pi^0\nu$	3·4	253
						$e^{\pm}\pi^0\nu$	5·0	358
K^0	0^-	± 1	$\frac{1}{2}$	497·8	50 per cent K short 50 per cent K long			
K^0_s	0^-		$\frac{1}{2}$	497·8	$0\cdot86\times10^{-10}$	$\pi^+\pi^-$	68·4	219
		± 1				$\pi^0\pi^0$	31·6	228
K^0_L	0^-		$\frac{1}{2}$	497·8	$5\cdot3\times10^{-8}$	$\pi^0\pi^0\pi^0$	25·5	93
						$\pi^+\pi^-\pi^0$	12·1	84
						$\pi^{\pm}\mu^{\mp}\nu$	27·3	253
						$\pi^{\pm}e^{\mp}\nu$	35·2	358
η^0	0^-	0	0	548·8	$\sim 3\times10^{-19}$ s (uncertainty principle)	$\gamma\gamma$	38·1	549
						$\pi^0\gamma\gamma$	3·4	414
						$\pi^0\pi^0\pi^0$	29·4	144
						$\pi^+\pi^-\pi^0$	23·4	135
						$\pi^+\pi^-\gamma$	5·5	269

B.2.2 *Baryons (B = 1)*

	J^P	S	I	Mass/ (MeV/c^2)	Mean life/s	Decays	Fraction /per cent	Q/MeV
p	$\frac{1}{2}^+$	0	$\frac{1}{2}$	938·3	stable			
n	$\frac{1}{2}^+$	0	$\frac{1}{2}$	939·6	$9\cdot6\times10^2$	$pe^-\nu$	100	1
Λ	$\frac{1}{2}^+$	-1	0	1115·6	$2\cdot5\times10^{-10}$	$p\pi^-$	65·3	38
						$n\pi^0$	34·7	41
Σ^+	$\frac{1}{2}^+$	-1	1	1189·4	$0\cdot8\times10^{-10}$	$p\pi^0$	52·8	116
						$n\pi^+$	47·2	110
Σ^0	$\frac{1}{2}^+$	-1	1	1192·6	$<1\cdot0\times10^{-14}$	$\Lambda\gamma$	100	77
Σ^-	$\frac{1}{2}^+$	-1	1	1197·3	$1\cdot6\times10^{-10}$	$n\pi^-$	100	118
Ξ^0	$\frac{1}{2}^+$	-2	$\frac{1}{2}$	1314·7	$3\cdot0\times10^{-10}$	$\Lambda\pi^0$	100	64
Ξ^-	$\frac{1}{2}^+$	-2	$\frac{1}{2}$	1321·3	$1\cdot6\times10^{-10}$	$\Lambda\pi^-$	100	66
Ω^-	$\frac{3}{2}^+$?	-3	0	1672·4	$1\cdot3\times10^{-10}$	$\Xi^0\pi^-$		217
						$\Xi^-\pi^0$		216
						ΛK^-		63

Baryon antiparticles are not separately listed.

References

ANDERSON, C. D., and NEDDERMEYER, S. H. (1936), *Physical Review*, vol. 50, pp. 263–71.

ANTIPOV, YU. M., BOLOTOV, V. N., DEVISHEV, M. I., DEVISHEVA, M. N., ISAKOV, V. V., KARPOV, I. I., KHODYREV, YU. S., KRENDELEV, V. A., LANDSBERG, L. C., LAPSHIN, V. G., LEBEDEV, A. A., MOROZOV, A. G., PROKOSHKIN, YU. D., RYBAKOV, V. A., RYKALIN, V. I., SAMOJLOV, A. V., SENKO, V. A., VISHNEVSKY, N. K., YETCH, F. A., and ZAJTZEV, A. M. (1969), *Physics Letters*, vol. 30B, pp. 576–80.

ASTBURY, P., FINOCCHIARO, G., MICHELINI, A., VERKERK, C., WEBSDALE, D., WEST, C., BEUSCH, W., GOBBI, B., PEPIN, M., POUCHON, M., and POLGAR, E. (1964, *Proceedings of XIIth International Conference on High Energy Physics,* Dubna, pp. 702–4.

BALDO-CEOLIN, M., BONETTI, A., GREENING, S., LIMENTANI, S., MERLIN, M., and VANDERHAEGHE, G. (1957), *Nuovo Cimento*, vol. 6, pp. 84–97.

BALTAY, C., FRANZINI, P., KIM, J., KIRSCH, L., ZANELLO, D., LEE-FRANZINI, J., LOVELESS, R., MCFADYEN, J., and YARGER, H. (1966), *Physical Review Letters*, vol. 16, pp. 1224–8.

BARKAS, W. H., DYER, J. N., and HECKMANN, H. H. (1963), *Physical Review Letters*, vol. 11, pp. 26–8.

BARNES, V. E., CULWICK, B. B., GUIDONI, P., KALBFLEISCH, G. R., LONDON, G. W., PALMER, R. B., RADOJICIC, D., RAHM, D. C., RAU, R. R., RICHARDSON, C. R., SAMIOS, N. P., and SMITH, J. R. (1965), *Physical Review Letters*, vol. 15, pp. 322–5.

BECCI, C., GALLINARO, G., and MORPURGO, G. (1965), *Nuovo Cimento*, vol. 39, pp. 409–12.

BELLETTINI, G., BEMPORAD, C., and BRACCINI, P. L. (1965), *Nuovo Cimento*, vol. 40A, pp. 1139–70.

BEMPORAD, C., BRACCINI, P. L., FOA, L., LÜBELSMEYER, L., and SCHMITZ, D. (1967), *Physics Letters,* vol. 25B, pp. 380–84.

BENZ, H., CHIKOVANI, G. E., DAMGAARD, G., FOCCACI, M. N., KIENZLE, W., LECHANOINE, C., MARTIN, M., NEF, C., SCHÜBELIN, P., BAUD, R., BOSNJAKOVICE, B., COTTERON, J., KLANNER, R., and WEITSCH, A. (1968), *Paper 240 to XIVth International Conference on High Energy Physics*, Vienna.

BERTANZA, L., BRISSON, V., CONNOLLY, P. L., HART, E. L., MITTRA, I. S., MONETI, G. C., RAU, R. R., SAMIOS, N. P., SKILLICORN, I. O., YAMAMOTO, S. S., GOLDBERG, M., GRAY, L., LEITNER, J., LICHTMAN, S., and WESTGARD, J. (1962), *Physical Review Letters*, vol. 9, pp. 180-83.

BJORKLUND, R., CRANDALL, W. E., MOYER, B. J., and YORK, H. F. (1950), *Physical Review*, vol. 77, pp. 213-18.

BLAND, R. W., BOWLER, M. G., BROWN, J. L., KADYK, J. A., GOLDHABER, G., GOLDHABER, S., SEEGER, V. H., and TRILLING, G. H. (1969), *Nuclear Physics*, vol. B13, pp. 595-621.

BLATON, J. (1950), *Kongelige Danske Videnskabernes Selskab, Mathematisk-Fysike*, Meddelelser, vol. 24, pp. 6-23.

BLIEDEN, H. R., FREYTAG, D., GEIBEL, J., HASSAN, A. R. F., KIENZLE, W., LEFEBRES, F., LEVRAT, B., MAGLIC, B. C., SEQUINOT, J., and SMITH, A. J. (1965), *Physics Letters*, vol. 19, pp. 444-8.

BLUM, W., BRANDT, S., COCCONI, V. T., CZYZEWSKI, O., DANYSZ, J., JOBES, M., KELLNER, G., MILLER, D., MORRISON, D. R. O., NEALE, W., and RUSHBROOKE, J. G. (1964), *Physical Review Letters*. vol. 13, p. 353.

BOWEN, R. A., CNOPS, A. M., FINOCCHIARO, G., MITTNER, P., DUFEY, J. P., GOBBI, B., POUCHON, M. A., and MULLER, A. (1967), *Physics Letters*, vol. 24B, pp. 207-8.

BROOKHAVEN NATIONAL LABORATORY (1964), *Physical Review Letters*, vol. 12, pp. 204-6.

CARLSON, A. G., HOOPER, J. E., and KING, D. T. (1950), *Philosophical Magazine*, vol. 41, pp. 701-24.

CARTWRIGHT, W. F., RICHMAN, C., WHITEHEAD, M. N., and WILCOX, H. A. (1953), *Physical Review*, vol. 91, pp. 677-88.

CHAMBERLAIN, O. (1960), *Annual Review of Nuclear Science*, vol. 10, pp. 161-91.

CHAMBERLAIN, O., SEGRE, E., WIEGAND, C., and YPSILANTIS, T. (1955), *Physical Review*, vol. 100, pp. 947-50.

CHRISTENSON, J. H., CRONIN, J. W., FITCH, V. L., and TURLAY, R. (1964), *Physical Review Letters*, vol. 13, pp. 138-40.

CNOPS, A. M., FINOCCHIARO, G., LASSALLE, J. C., MITTNER, P., ZANELLA, P., DUFEY, J. P., GOBBI, B., POUCHON, M. A., and MULLER, A. (1966), *Physics Letters*, vol. 22, pp. 546-50.

CONVERSI, M., PANCINI, E., and PICCIONI, O. (1947), *Physical Review*, vol. 71, pp. 209-10.

CRENNELL, D. J., KARSHON, U., LAI, K. W., SCARR, J. M., and SKILLICORN, I. O. (1968), *Physics Letters*, vol. 28B, pp. 136-9.

DALITZ, R. H. (1967), *Proceedings of XIIIth International Conference on High Energy Physics*, Berkeley, 1966, pp. 215-34.

DANBY, G., GAILLARD, J.-M., GOULIANOS, K., LEDERMAN, L. M., MISTRY, N., SCHWARTZ, M., and STEINBERGER, J. (1962), *Physical Review Letters*, vol. 9, pp. 36-44.

DURBIN, R., LOAR, H., and STEINBERGER, J. (1951), *Physical Review*, vol. 83, pp. 646-8, and vol. 84, pp. 581-2.

FICKINGER, W. J., PICKUP, E., ROBINSON, D. K., and SALANT, E. O. (1962), *Physical Review*, vol. 125, pp. 2082–90.

FOCACCI, M. N., KIENZLE, W., LEVRAT, B., MAGLIC, B. C., and MARTIN, M. (1966), *Physical Review Letters*, vol. 17, pp. 890–93.

FOWLER, E. C. (1966), *Bulletin of American Physical Society*, vol. 11, p. 380.

FRANZINETTI, C., and MORPURGO, G. (1957), *Supplemento Nuovo Cimento*, vol. 2, pp. 535–6.

GALLINARO, G., and MORPURGO, D. (1966), *Physics Letters*, vol. 23, pp. 609–13.

GELL-MANN, M. (1953), *Physical Review*, vol. 92, pp. 833–4.

GELL-MANN, M. (1962), *Physical Review*, vol. 125, pp. 1067–84.

GELL-MANN, M., and PAIS, A. (1954), *Proceedings of the Glasgow Conference on Nuclear and Meson Physics*, pp. 342–52.

GIBSON, W. M. (1971), *Nuclear Reactions*, Penguin.

HAGEDORN, R. (1963), *Relativistic Kinematics*, Benjamin.

HENDERSON, C. (1970), *Cloud and Bubble Chambers*, Methuen.

JAUCH, J. M. (1959), *Strange Particle Physics*, CERN report no. 59–35, pp. 19–37.

JONES, W. G., BINNIE, D. M., DUANE, A., HORSEY, J. P., MASON, D. C., NEWTH, J. A., RAHMAN, I. U., WALTERS, J., HORWITZ, N., and PALIT, P. (1966), *Physics Letters*, vol. 23, pp. 597–600.

KIENZLE, W. (1968), CERN internal report no. 68–25, Figs. 45, 46, 65.

LANDÉ, K., BOOTH, E. T., IMPEDUGLIA, J., LEDERMAN, L. M., and CHINOWSKI, W. (1956), *Physical Review*, vol. 103, pp. 1901–4.

LARRIBE, A., LEVEQUE, A., MULLER, A., PAULI, E., REVEL, D., TALLINI, T., LITCHFIELD, P. J., RANGAN, L. K., SEGAR, A. M., SMITH, J. R., FINNEY, P. J., FISHER, C. M., and PICKUP, E. (1966), *Physics Letters,* vol. 23, pp. 600–64.

LATTES, C. M. G., OCCHIALINI, G. P. S., and POWELL, C. F. (1947), *Nature*, vol. 160, pp. 453–6.

LEE, T. D., and YANG, C. N. (1956), *Physical Review*, vol. 104, pp. 254–8.

LEPRINCE-RINGUET, L., and LHERITIER, M. (1944), *Comptes Rendus de L'Académie des Sciences*, vol. 219, pp. 618–20.

LINDENBAUM, S. J., and YUAN, L. C. L. (1961), *Methods of Experimental Physics*, vol. 5, part A, pp. 162–94.

LIPKIN, H. J. (1965), *Lie Groups for Pedestrians*, North Holland.

LÜDERS, G. (1954), *Kongelige Danske Videnskabarnes Selskab, Mathematisk-Fysike*, Meddelelser, vol. 28, no. 5.

MCCUSKER, C. B. A., and CAIRNS, I. (1969), *Physical Review Letters*, vol. 23, pp. 658–9.

MUIRHEAD, H. (1965), *The Physics of Elementary Particles*, Pergamon.

NE'EMAN, Y. (1961), *Nuclear Physics*, vol. 26, pp. 222–9.

NISHIJIMA, K. (1954), *Progress of Theoretical Physics*, vol. 13, pp. 285–304.

O'CEALLAIGH, C. (1950), *Philosophical Magazine*, vol. 41, pp. 838–48.

PAULI, W. (1955), *Niels Bohr and the Development of Physics*, McGraw-Hill.

PETERSON, J. R. (1957), *Physical Review*, vol. 105, pp. 693–706.

PEVSNER, A., KRAEMER, R., NUSSBAUM, M., RICHARDSON, C., SCHLEIN, P., STRAND, R., TOOHIG, T., BLOCK, M., ENGLER, A., GESSAROLI, R., and

MELTZER, C. (1961), *Physical Review Letters*, vol. 7, pp. 421-3.

PJERROU, G. M., PROWSE, D. J., SCHLEIN, P., SLATER, W. E., STORK, D. H., and TICHO, H. K. (1962), *Physical Review Letters*, vol. 9, pp. 114-17.

PLANO, R., PRODELL, A., SAMIOS, N., SCHWARTZ, M., and STEINBERGER, J. (1959), *Physical Review Letters*, vol. 3, pp. 525-7.

POWELL, C., FOWLER, P. H., and PERKINS, D. (1959), *The Study of Elementary Particles by the Photographic Method*, Pergamon.

ROCHESTER, G. D., and Butler, C. C. (1947), *Nature*, vol. 160, pp. 855-7.

SAKATA, S. (1956), *Progress of Theoretical Physics*, vol. 16, pp. 686-8.

SAKURAI, J. J. (1962), *Physical Review Letters*, vol. 9, pp. 472-5.

SANDWEISS, J. (1967), in R. P. Shutt (ed.), *Bubble and Spark Chambers*, vol. 2, pp. 195-287, Academic Press.

SHUTT, R. P. (ed.) (1967), *Bubble and Spark Chambers*, vols. 1 and 2, Academic Press.

STERNHEIMER, R. M., and CORK, B. (1963), in Yuan, L. C. L., and Wu, C. (eds.), *Methods of Experimental Physics*, Vol. B, pp. 691-760, Academic Press.

THOMPSON, R. W., COHN, H. O., and FLUM, R. S. (1951), *Physical Review*, vol. 83, p. 175.

WALKER, W. D., CARROLL, A. S., GARFINKEL, A., and OH, B. Y. (1967), *Physical Review Letters*, vol. 18, pp. 630-33.

WATSON, M. B., FERRO-LUZZI, M., and TRIPP, R. D. (1963), *Physical Review*, vol. 131, pp. 2248-81.

WU, C. S., HAYWARD, R. W., HOPPES, D. D., and HUDSON, R. P. (1957), *Physical Review*, vol. 105, pp. 1413-15.

YUKAWA, H. (1935), *Proceedings of the Physico-Mathematical Society of Japan* vol. 17, pp. 48-57.

Index